An Introduction to Modern Cosmology

Second Edition

An Introduction To Modern Cosmology

Second Edition

Andrew Liddle
University of Sussex, UK

NORWICH CITY COLLEGE LIBRARY

Stock No.	217426		
Class	523·1 LID		
Cat.	SSA	Proc.	3WL

WILEY

Copyright © 2003 John Wiley & Sons Ltd, The Atrium, Southern Gate, Chichester,
West Sussex PO19 8SQ, England

Telephone (+44) 1243 779777

Email (for orders and customer service enquiries): cs-books@wiley.co.uk
Visit our Home Page on www.wileyeurope.com or www.wiley.com

Reprinted April and December 2004, December 2005, October 2007
Reprinted with corrections February 2007, December 2008

All Rights Reserved. No part of this publication may be reproduced, stored in a retrieval system or transmitted in any form or by any means, electronic, mechanical, photocopying, recording, scanning or otherwise, except under the terms of the Copyright, Designs and Patents Act 1988 or under the terms of a licence issued by the Copyright Licensing Agency Ltd, 90 Tottenham Court Road, London W1T 4LP, UK, without the permission in writing of the Publisher. Requests to the Publisher should be addressed to the Permissions Department, John Wiley & Sons Ltd, The Atrium, Southern Gate, Chichester, West Sussex PO19 8SQ, England, or emailed to permreq@wiley.co.uk, or faxed to (+44) 1243 770571.

This publication is designed to provide accurate and authoritative information in regard to the subject matter covered. It is sold on the understanding that the Publisher is not engaged in rendering professional services. If professional advice or other expert assistance is required, the services of a competent professional should be sought.

Other Wiley Editorial Offices

John Wiley & Sons Inc., 111 River Street, Hoboken, NJ 07030, USA

Jossey-Bass, 989 Market Street, San Francisco, CA 94103-1741, USA

Wiley-VCH Verlag GmbH, Boschstr. 12, D-69469 Weinheim, Germany

John Wiley & Sons Australia Ltd, 33 Park Road, Milton, Queensland 4064, Australia

John Wiley & Sons (Asia) Pte Ltd, 2 Clementi Loop #02-01, Jin Xing Distripark, Singapore 129809

John Wiley & Sons Canada Ltd, 22 Worcester Road, Etobicoke, Ontario, Canada M9W 1L1

Wiley also publishes in books in a variety of electronic formats. Some content that
appears in print may not be available in electronic books.

British Library Cataloging in Publication Data

A catalogue record for this book is available from the British Library

ISBN 13: 978-0-470-84835-7 (P/B)
ISBN 13: 978-0-470-84834-0 (H/B)

Produced from author's LaTeX files
Printed and bound in Great Britain by CPI Antony Rowe, Chippenham, Wilts

To my grandmothers

Contents

Preface

The development of cosmology will no doubt be seen as one of the scientific triumphs of the twentieth century. At its beginning, cosmology hardly existed as a scientific discipline. By its end, the Hot Big Bang cosmology stood secure as the accepted description of the Universe as a whole. Telescopes such as the Hubble Space Telescope are capable of seeing light from galaxies so distant that the light has been travelling towards us for most of the lifetime of the Universe. The cosmic microwave background, a fossil relic of a time when the Universe was both denser and hotter, is routinely detected and its properties examined. That our Universe is presently expanding is established without doubt.

We are presently in an era where understanding of cosmology is shifting from the qualitative to the quantitative, as rapidly-improving observational technology drives our knowledge forward. The turn of the millennium saw the establishment of what has come to be known as the Standard Cosmological Model, representing an almost universal consensus amongst cosmologists as to the best description of our Universe. Nevertheless, it is a model with a major surprise — the belief that our Universe is presently experiencing accelerated expansion. Add to that ongoing mysteries such as the properties of the so-called dark matter, which is believed to be the dominant form of matter in the Universe, and it is clear that we have some way to go before we can say that a full picture of the physics of the Universe is in our grasp.

Such a bold endeavour as cosmology easily captures the imagination, and over recent years there has been increasing demand for cosmology to be taught at university in an accessible manner. Traditionally, cosmology was taught, as it was to me, as the tail end of a general relativity course, with a derivation of the metric for an expanding Universe and a few solutions. Such a course fails to capture the flavour of modern cosmology, which takes classic physical sciences like thermodynamics, atomic physics and gravitation and applies them on a grand scale.

In fact, introductory modern cosmology can be tackled in a different way, by avoiding general relativity altogether. By a lucky chance, and a subtle bit of cheating, the correct equations describing an expanding Universe can be obtained from Newtonian gravity. From this basis, one can study all the triumphs of the Hot Big Bang cosmology — the expansion of the Universe, the prediction of its age, the existence of the cosmic microwave background, and the abundances of light elements such as helium and deuterium — and even go on to discuss more speculative ideas such as the inflationary cosmology.

The origin of this book, first published in 1998, is a short lecture course at the University of Sussex, around 20 lectures, taught to students in the final year of a bachelor's degree or the penultimate year of a master's degree. The prerequisites are all very standard physics, and the emphasis is aimed at physical intuition rather than mathematical rigour. Since the book's publication cosmology has moved on apace, and I have also become aware of the need for a somewhat more extensive range of material, hence this second edition. To summarize the differences from the first edition, there is more stuff than before, and the stuff that was already there is now less out-of-date.

Cosmology is an interesting course to teach, as it is not like most of the other subjects taught in undergraduate physics courses. There is no perceived wisdom, built up over a century or more, which provides an unquestionable foundation, as in thermodynamics, electromagnetism, and even quantum mechanics and general relativity. Within our broad-brush picture the details often remain rather blurred, changing as we learn more about the Universe in which we live. Opportunities crop up during the course to discuss new results which impact on cosmologists' views of the Universe, and for the lecturer to impose their own prejudices on the interpretation of the ever-changing observational situation. Unless I've changed jobs (in which case I'm sure `www.google.com` will hunt me down), you can follow my own current prejudices by checking out this book's WWW Home Page at

`http://astronomy.susx.ac.uk/~andrewl/cosbook.html`

There you can find some updates on observations, and also a list of any errors in the book that I am aware of. If you are confident you've found one yourself, and it's not on the list, I'd be very pleased to hear of it.

The structure of the book is a central 'spine', the main chapters from one to fifteen, which provide a self-contained introduction to modern cosmology more or less reproducing the coverage of my Sussex course. In addition there are six Advanced Topic chapters, each with prerequisites, which can be added to extend the course as desired. Ordinarily the best time to tackle those Advanced Topics is immediately after their prerequisites have been attained, though they could also be included at any later stage.

I'm extremely grateful to the reviewers of the original draft manuscript, namely Steve Eales, Coel Hellier and Linda Smith, for numerous detailed comments which led to the first edition being much better than it would have otherwise been. Thanks also to those who sent me useful comments on the first edition, in particular Paddy Leahy and Michael Rowan-Robinson, and of course to all the Wiley staff who contributed. Matthew Colless and Michael Turner provided two of the figures, and Martin Hendry, Martin Kunz and Franz Schunck helped with three others, while two figures were generated from NASA's *SkyView* facility (`http://skyview.gsfc.nasa.gov`) located at the NASA Goddard Space Flight Center. A library of images, including full-colour versions of several images reproduced here in black and white to keep production costs down, can be found via the book's Home Page as given above.

Andrew R Liddle
Brighton
February 2003

Some fundamental constants

Newton's constant	G		$6.672 \times 10^{-11}\,\mathrm{m^3\,kg^{-1}\,sec^{-2}}$
Speed of light	c		$2.998 \times 10^{8}\,\mathrm{m\,sec^{-1}}$
		or	$3.076 \times 10^{-7}\,\mathrm{Mpc\,yr^{-1}}$
Reduced Planck constant	$\hbar = h/2\pi$		$1.055 \times 10^{-34}\,\mathrm{m^2\,kg\,sec^{-1}}$
Boltzmann constant	k_B		$1.381 \times 10^{-23}\,\mathrm{J\,K^{-1}}$
		or	$8.619 \times 10^{-5}\,\mathrm{eV\,K^{-1}}$
Radiation constant	$\alpha = \pi^2 k_\mathrm{B}^4/15\hbar^3 c^3$		$7.565 \times 10^{-16}\,\mathrm{J\,m^{-3}\,K^{-4}}$
Electron mass–energy	$m_\mathrm{e}\,c^2$		$0.511\,\mathrm{MeV}$
Proton mass–energy	$m_\mathrm{p}\,c^2$		$938.3\,\mathrm{MeV}$
Neutron mass–energy	$m_\mathrm{n}\,c^2$		$939.6\,\mathrm{MeV}$
Thomson cross-section	σ_e		$6.652 \times 10^{-29}\,\mathrm{m^2}$
Free neutron half-life	t_half		$614\,\mathrm{sec}$

Some conversion factors

$1\,\mathrm{pc} = 3.261\text{ light years} = 3.086 \times 10^{16}\,\mathrm{m}$

$1\,\mathrm{yr} = 3.156 \times 10^{7}\,\mathrm{sec}$

$1\,\mathrm{eV} = 1.602 \times 10^{-19}\,\mathrm{J}$

$1\,M_\odot = 1.989 \times 10^{30}\,\mathrm{kg}$

$1\,\mathrm{J} = 1\,\mathrm{kg\,m^2\,sec^{-2}}$

$1\,\mathrm{Hz} = 1\,\mathrm{sec^{-1}}$

Commonly-used symbols

z	redshift	*defined on page* 9, 35
H_0	Hubble constant	9, 45
r	physical distance	9
v	velocity	9
f	frequency	12
T	temperature	13
k_B	Boltzmann constant	13
ϵ	energy density	15
α	radiation constant	15
G	Newton's gravitational constant	17
ρ	mass density	18
a	scale factor	19
x	comoving distance	19
k	curvature	20
p	pressure	22
	(or occasionally momentum	11)
H	Hubble parameter	34
n, N	number density	39
h	Hubble constant	46
	(or Planck's constant	12)
Ω_0	present density parameter	47
ρ_c	critical density	47
Ω	density parameter	48
Ω_k	curvature 'density parameter'	48
q_0	deceleration parameter	48
Λ	cosmological constant	51
Ω_Λ	cosmological constant density parameter	52
t	time	57
t_0	present age	57
Ω_B	baryon density parameter	64
Y_4	helium abundance	93
d_{lum}	luminosity distance	129
d_{diam}	angular diameter distance	132
$\Delta T/T$, C_ℓ	cosmic microwave background anisotropies	152, 153

Chapter 1

A Brief History of Cosmological Ideas

The cornerstone of modern cosmology is the belief that the place which we occupy in the Universe is in no way special. This is known as the **cosmological principle**, and it is an idea which is both powerful and simple. It is intriguing, then, that for the bulk of the history of civilization it was believed that we occupy a very special location, usually the centre, in the scheme of things.

The ancient Greeks, in a model further developed by the Alexandrian Ptolemy, believed that the Earth must lie at the centre of the cosmos. It would be circled by the Moon, the Sun and the planets, and then the 'fixed' stars would be yet further away. A complex combination of circular motions, Ptolemy's Epicycles, was devised in order to explain the motions of the planets, especially the phenomenon of retrograde motion where planets appear to temporarily reverse their direction of motion. It was not until the early 1500s that Copernicus stated forcefully the view, initiated nearly two thousand years before by Aristarchus, that one should regard the Earth, and the other planets, as going around the Sun. By ensuring that the planets moved at different speeds, retrograde motion could easily be explained by this theory. However, although Copernicus is credited with removing the anthropocentric view of the Universe, which placed the Earth at its centre, he in fact believed that the Sun was at the centre.

Newton's theory of gravity put what had been an empirical science (Kepler's discovery that the planets moved on elliptical orbits) on a solid footing, and it appears that Newton believed that the stars were also suns pretty much like our own, distributed evenly throughout infinite space, in a static configuration. However it seems that Newton was aware that such a static configuration is unstable.

Over the next two hundred years, it became increasingly understood that the nearby stars are not evenly distributed, but rather are located in a disk-shaped assembly which we now know as the Milky Way galaxy. The Herschels were able to identify the disk structure in the late 1700s, but their observations were not perfect and they wrongly concluded that the solar system lay at its centre. Only in the early 1900s was this convincingly overturned, by Shapley, who realized that we are some two-thirds of the radius away from the centre of the galaxy. Even then, he apparently still believed our galaxy to be at the centre of the

Universe. Only in 1952 was it finally demonstrated, by Baade, that the Milky Way is a fairly typical galaxy, leading to the modern view, known as the **cosmological principle** (or sometimes the Copernican principle) that the Universe looks the same whoever and wherever you are.

It is important to stress that the cosmological principle isn't exact. For example, no one thinks that sitting in a lecture theatre is exactly the same as sitting in a bar, and the interior of the Sun is a very different environment from the interstellar regions. Rather, it is an approximation which we believe holds better and better the larger the length scales we consider. Even on the scale of individual galaxies it is not very good, but once we take very large regions (though still much smaller than the Universe itself), containing say a million galaxies, we expect every such region to look more or less like every other one. The cosmological principle is therefore a property of the global Universe, breaking down if one looks at local phenomena.

The cosmological principle is the basis of the Big Bang Cosmology. The Big Bang is the best description we have of our Universe, and the aim of this book is to explain why. The Big Bang is a picture of our Universe as an evolving entity, which was very different in the past as compared to the present. Originally, it was forced to compete with a rival idea, the Steady State Universe, which holds that the Universe does not evolve but rather has looked the same forever, with new material being created to fill the gaps as the Universe expands. However, the observations I will describe now support the Big Bang so strongly that the Steady State theory is almost never considered.

Chapter 2

Observational Overview

For most of history, astronomers have had to rely on light in the visible part of the spectrum in order to study the Universe. One of the great astronomical achievements of the 20th century was the exploitation of the full electromagnetic spectrum for astronomical measurements. We now have instruments capable of making observations of radio waves, microwaves, infrared light, visible light, ultraviolet light, X-rays and gamma rays, which all correspond to light waves of different (in this case increasing) frequency. We are even entering an epoch where we can go beyond the electromagnetic spectrum and receive information of other types. A remarkable feature of observations of a nearby supernova in 1987 was that it was also seen through detection of neutrinos, an extraordinarily weakly interacting type of particle normally associated with radioactive decay. Very high energy cosmic rays, consisting of highly-relativistic elementary particles, are now routinely detected, though as yet there is no clear understanding of their astronomical origin. And as I write, experiments are starting up with the aim of detecting gravitational waves, ripples in space-time itself, and ultimately of using them to observe astronomical events such as colliding stars.

The advent of large ground-based and satellite-based telescopes operating in all parts of the electromagnetic spectrum has revolutionized our picture of the Universe. While there are probably gaps in our knowledge, some of which may be important for all we know, we do seem to have a consistent picture, based on the cosmological principle, of how material is distributed in the Universe. My discussion here is brief; for a much more detailed discussion of the observed Universe, see Rowan-Robinson's book 'Cosmology' (full reference in the Bibliography). A set of images, including full-colour versions of the figures in this chapter, can be found via the book's Home Page as given in the Preface.

2.1 In visible light

Historically, our picture of the Universe was built up through ever more careful observations using visible light.

Stars: The main source of visible light in the Universe is nuclear fusion within stars. The Sun is a fairly typical star, with a mass of about 2×10^{30} kilograms. This is known as a solar mass, indicated $M_\odot$, and is a convenient unit for measuring masses. The

Figure 2.1 If viewed from above the disk, our own Milky Way galaxy would probably resemble the M100 galaxy, imaged here by the Hubble Space telescope. [Figure courtesy NASA]

nearest stars to us are a few light years away, a light year being the distance (about 10^{16} metres) that light can travel in a year. For historical reasons, an alternative unit, known as the **parsec** and denoted 'pc',[1] is more commonly used in cosmology. A parsec equals 3.261 light years. In cosmology, one seldom considers individual stars, instead preferring to adopt as the smallest considered unit the conglomerations of stars known as ...

Galaxies: Our solar system lies some way off-centre in a giant disk structure known as the Milky Way galaxy. It contains a staggering hundred thousand million (10^{11}) or so stars, with masses ranging from about a tenth that of our Sun to tens of times larger. It consists of a central bulge, plus a disk of radius 12.5 kiloparsecs (kpc, equal to 10^3 pc) and a thickness of only about 0.3 kpc. We are located in the disk about 8 kpc from the centre. The disk rotates slowly (and also differentially, with the outer edges moving more slowly, just as more distant planets in the solar system orbit more slowly). At our radius, the galaxy rotates with a period of 200 million years. Because we are within it, we can't get an image of our own galaxy, but it probably looks not unlike the M100 galaxy shown in Figure 2.1.

Our galaxy is surrounded by smaller collections of stars, known as globular clusters. These are distributed more or less symmetrically about the bulge, at distances of 5–

[1] A parsec is defined as the distance at which the mean distance between the Earth and Sun subtends a second of arc. The mean Earth–Sun distance (called an Astronomical Unit) is 1.496×10^{11} m, and dividing that by $\tan(1\,\mathrm{arcsec})$ gives $1\,\mathrm{pc} = 3.086 \times 10^{16}$ m.

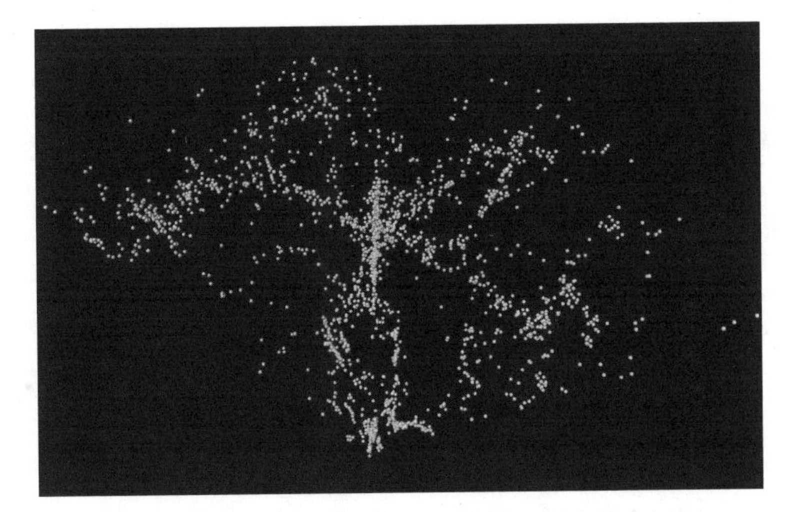

Figure 2.2 A map of galaxy positions in a narrow slice of the Universe, as identified by the CfA (Center for Astrophysics) redshift survey. Our galaxy is located at the apex, and the radius is around 200 Mpc. The galaxy positions were obtained by measurement of the shift of spectral lines, as described in Section 2.4. While more modern and extensive galaxy redshift surveys exist, this survey still gives one of the best impressions of structure in the Universe. [Figure courtesy Lars Christensen]

30 kpc. Typically they contain a million stars, and are thought to be remnants of the formation of the galaxy. As we shall discuss later, it is believed that the entire disk and globular cluster system may be embedded in a larger spherical structure known as the galactic halo.

Galaxies are the most visually striking and beautiful astronomical objects in the Universe, exhibiting a wide range of properties. However, in cosmology the detailed structure of a galaxy is usually irrelevant, and galaxies are normally thought of as point-like objects emitting light, often broken into sub-classes according to colours, luminosities and morphologies.

The local group: Our galaxy resides within a small concentrated group of galaxies known as the local group. The nearest galaxy is a small irregular galaxy known as the Large Magellanic Cloud (LMC), which is 50 kpc away from the Sun. The nearest galaxy of similar size to our own is the Andromeda Galaxy, at a distance of 770 kpc. The Milky Way is one of the largest galaxies in the local group. A typical galaxy group occupies a volume of a few cubic **megaparsecs**. The megaparsec, denoted Mpc and equal to a million parsecs, is the cosmologist's favourite unit for measuring distances, because it is roughly the separation between neighbouring galaxies. It equals 3.086×10^{22} metres.

Clusters of galaxies, superclusters and voids: Surveying larger regions of the Universe, say on a scale of 100 Mpc, one sees a variety of large-scale structures, as shown in Figure 2.2. This figure is not a photograph, but rather a carefully constructed map of the nearby region of our Universe, on a scale of about $1{:}10^{27}$! In some

Figure 2.3 Images of the Coma cluster of galaxies in visible light (left) and in X-rays (right), on the same scale. Colour versions can be found on the book's WWW site. [Figures courtesy of the Digitized Sky Survey, ROSAT and *SkyView*]

places galaxies are clearly grouped into clusters of galaxies; a famous example is the Coma cluster of galaxies. It is about 100 Mpc away from our own galaxy, and appears in Figure 2.2 as the dense region in the centre of the map. The left panel of Figure 2.3 shows an optical telescope image of Coma; although the image resembles a star field, each point is a distinct galaxy. Coma contains perhaps 10 000 galaxies, mostly too faint to show in this image, orbitting in their common gravitational field. However, most galaxies, sometimes called field galaxies, are not part of a cluster. Galaxy clusters are the largest gravitationally-collapsed objects in the Universe, and they themselves are grouped into superclusters, perhaps joined by filaments and walls of galaxies. In between this 'foamlike' structure lie large voids, some as large as 50 Mpc across. Structures in the Universe will be further described in Advanced Topic 5.

Large-scale smoothness: Only once we get to even larger scales, hundreds of megaparsecs or more, does the Universe begin to appear smooth. Recent extremely large galaxy surveys, the 2dF galaxy redshift survey and the Sloan Digital Sky Survey, have surveyed volumes around one hundred times the size of the CfA survey, each containing hundreds of thousands of galaxies. Such surveys do not find any huge structures on scales greater than those seen in the CfA survey; the galaxy superclusters and voids just discussed are likely to be the biggest structures in the present Universe.

The belief that the Universe does indeed become smooth on the largest scales, the cosmological principle, is the underpinning of modern cosmology. It is interesting that while the smoothness of the matter distribution on large scales has been a key assumption of cosmology for decades now, it is only fairly recently that it has been possible to provide a convincing observational demonstration.

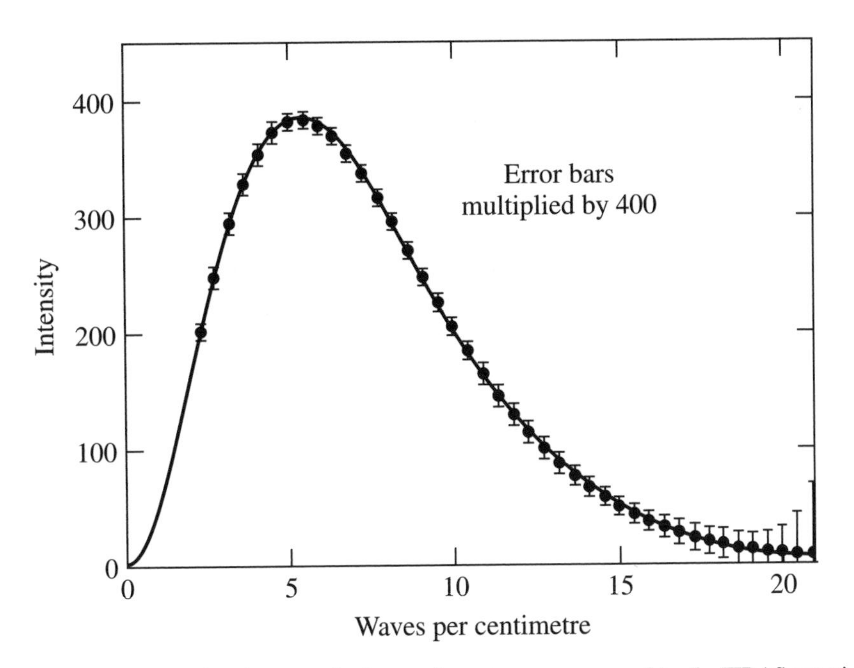

Figure 2.4 The cosmic microwave background spectrum as measured by the FIRAS experiment on the COBE satellite. The error bars are so small that they have been multiplied by 400 to make them visible on this plot, and the best-fit black-body spectrum at $T = 2.725$ Kelvin, shown by the line, is an excellent fit.

2.2 In other wavebands

Observations using visible light provide us with a good picture of what's going on in the present-day Universe. However, many other wavebands make vital contributions to our understanding.

Microwaves: For cosmology, this is by far the most important waveband. Penzias & Wilson's accidental discovery in 1965 that the Earth is bathed in microwave radiation, with a black-body spectrum at a temperature of around 3 Kelvin, was and is one of the most powerful pieces of information in support of the Big Bang theory, around which cosmology is now based. Observations by the FIRAS (Far InfraRed Absolute Spectrometer) experiment on board the COBE (COsmic Background Explorer) satellite have confirmed that the radiation is extremely close to the black-body form at a temperature 2.725 ± 0.001 Kelvin. This data is shown in Figure 2.4. Furthermore, the temperature coming from different parts of the sky is astonishingly uniform, and this presents the best evidence that we can use the cosmological principle as the foundation of cosmology. In fact, it has recently been possible to identify tiny variations, only one part in a hundred thousand, between the intensities of the microwaves coming from different directions. It is believed that these are intimately related to the origin of structure in the Universe. This fascinating topic is revolutionizing cosmology, and will be explored further in Advanced Topic 5.

Radio waves: A powerful way of gaining high-resolution maps of very distant galaxies is by mapping in the radio part of the spectrum. Many of the furthest galaxies known were detected in this way.

Infrared: Carrying out surveys in the infrared part of the spectrum, as was done by the highly-successful IRAS (InfraRed Astronomical Satellite) in the 1980s, is an excellent way of spotting young galaxies, in which star formation is at an early stage. Infrared surveys pick up a somewhat different population of galaxies to surveys carried out in optical light, though obviously the brightest galaxies are seen in both. Infrared is particularly good for looking through the dust in our own galaxy to see distant objects, as it is absorbed and scattered much less strongly than visible radiation. Accordingly, it is best for studying the region close to our galactic plane, where obscuration by dust is strongest.

X-rays: These are a vital probe of clusters of galaxies; in between the galaxies lies gas so hot that it emits in the X-ray part of the spectrum, corresponding to a temperature of tens of millions of Kelvin. This gas is thought to be remnant material from the formation of the galaxies, which failed to collapse to form stars. X-ray emission from the Coma galaxy cluster is shown in the right panel of Figure 2.3. The individual galaxies seen in the visible light image in the left panel are almost all invisible in X-rays, with the bright diffuse X-ray emission from the hot gas dominating the image.

2.3 Homogeneity and isotropy

The evidence that the Universe becomes smooth on large scales supports the use of the cosmological principle. It is therefore believed that our large-scale Universe possesses two important properties, **homogeneity** and **isotropy**. Homogeneity is the statement that the Universe looks the same at each point, while isotropy states that the Universe looks the same in all directions.

These do not automatically imply one another. For example, a Universe with a uniform magnetic field is homogeneous, as all points are the same, but it fails to be isotropic because directions along the field lines can be distinguished from those perpendicular to them. Alternatively, a spherically-symmetric distribution, viewed from its central point, is isotropic but not necessarily homogeneous. However, if we require that a distribution is isotropic about *every* point, then that does enforce homogeneity as well.

As mentioned earlier, the cosmological principle is not exact, and so our Universe does not respect exact homogeneity and isotropy. Indeed, the study of departures from homogeneity is currently the most prominent research topic in cosmology. I'll introduce this in Advanced Topic 5, but in the main body of this book I am concerned only with the behaviour of the Universe as a whole, and so will be assuming large-scale homogeneity and isotropy.

2.4 The expansion of the Universe

A key piece of observational evidence in cosmology is that almost everything in the Universe appears to be moving away from us, and the further away something is, the more rapid its recession appears to be. These velocities are measured via the **redshift**, which is basically the Doppler effect applied to light waves. Galaxies have a set of absorption and emission lines identifiable in their spectra, whose characteristic frequencies are well known. However, if a galaxy is moving towards us, the light waves get crowded together, raising the frequency. Because blue light is at the high-frequency end of the visible spectrum, this is known as a blueshift. If the galaxy is receding, the characteristic lines move towards the red end of the spectrum and the effect is known as a redshift. This technique was first used to measure a galaxy's velocity by Vesto Slipher around 1912, and was applied systematically by one of the most famous cosmologists, Edwin Hubble, in the following decades.

It turns out that almost all galaxies are receding from us, so the standard terminology is redshift z, defined by

$$z = \frac{\lambda_{\rm obs} - \lambda_{\rm em}}{\lambda_{\rm em}}, \tag{2.1}$$

where $\lambda_{\rm em}$ and $\lambda_{\rm obs}$ are the wavelengths of light at the points of emission (the galaxy) and observation (us). If a nearby object is receding at a speed v, then its redshift is

$$z = \frac{v}{c}, \tag{2.2}$$

where c is the speed of light.[2] Figure 2.5 shows velocity against distance, a plot known as the Hubble diagram, for a sample of 1355 galaxies.

Hubble realized that his observations, which were of course much less extensive than those available to us now, showed that the velocity of recession was proportional to the distance of an object from us:

$$\vec{v} = H_0 \, \vec{r}. \tag{2.3}$$

This is known as **Hubble's law**, and the constant of proportionality H_0 is known as **Hubble's constant**. Hubble's law isn't exact, as the cosmological principle doesn't hold perfectly for nearby galaxies, which typically possess some random motions known as peculiar velocities. But it does describe the average behaviour of galaxies extremely well. Hubble's law gives the picture of our Universe illustrated in Figure 2.6, where the nearby galaxies have the smallest velocity relative to ours. Over the years many attempts have been made to find accurate values for the proportionality constant, but, as we will see in

[2]This formula ignores special relativity and so is valid only for speeds $v \ll c$. If you're interested, the special relativity result, of which this is an expansion for small v/c, is

$$1 + z = \sqrt{\frac{1 + v/c}{1 - v/c}}.$$

However, for distant objects in cosmology there are further considerations, concerning the propagation time of the light and how the relative velocity might change during it, and so this expression should not be used.

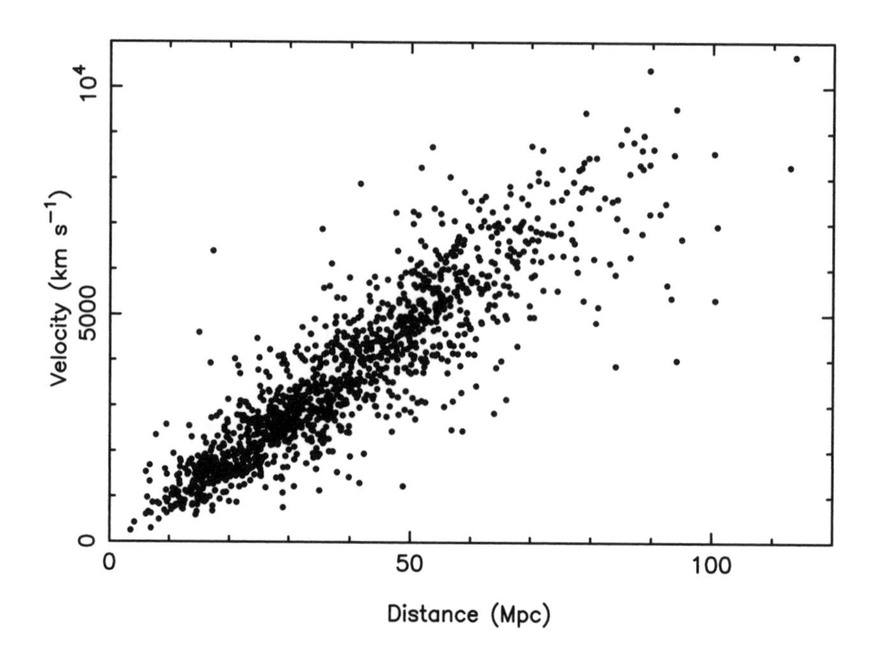

Figure 2.5 A plot of velocity versus estimated distance for a set of 1355 galaxies. A straight-line relation implies Hubble's law. The considerable scatter is due to observational uncertainties and random galaxy motions, but the best-fit line accurately gives Hubble's law. [The x-axis scale assumes a particular value of H_0.]

Chapter 6, a consensus is only now being reached.

At first sight, it seems that the cosmological principle must be violated if we observe everything to be moving away from us, since that apparently places us at the centre of the Universe. However, nothing could be further from the truth. In fact, *every* observer sees all objects rushing away from them with velocity proportional to distance. It is perhaps easiest to convince yourself of this by setting up a square grid with recession velocity proportional to distance from the central grid-point. Then transform the frame of reference to a nearby grid-point, and you'll find that the Hubble law still holds about the new 'centre'. This only works because of the linear relationship between velocity and distance; any other law and it wouldn't work.

So, although expanding, the Universe looks just the same whichever galaxy we choose to imagine ourselves within. A common analogy is to imagine baking a cake with raisins in it, or blowing up a balloon with dots on its surface. As the cake rises (or the balloon is inflated), the raisin (or dots) move apart. From each one, it seems that all the others are receding, and the further away they are the faster that recession is.

Because everything is flying away from everything else, we conclude that in the distant past everything in the Universe was much closer together. Indeed, trace the history back far enough and everything comes together. The initial explosion is known as the **Big Bang**, and a model of the evolution of the Universe from such a beginning is known as the **Big Bang Cosmology**. Later on, we will find out why it is commonly called the **Hot Big Bang**.

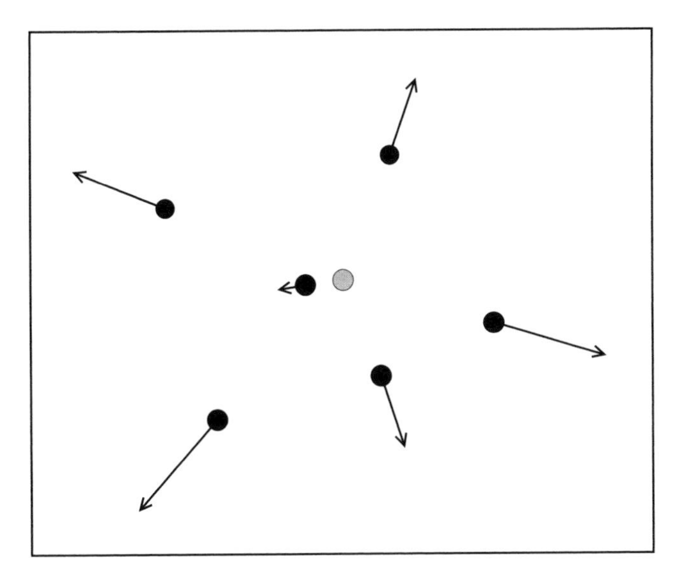

Figure 2.6 According to Hubble's law, the further away from us a galaxy is, the faster it is receding.

2.5 Particles in the Universe

2.5.1 What particles are there?

Everything in the Universe is made up of fundamental particles, and the behaviour of the Universe as a whole depends on the properties of these particles.

One crucial question is whether a particle is moving relativistically or not. Any particle has two contributions to its energy, one being the kinetic energy and the other being the mass–energy, which combine to give

$$E_{\rm total}^2 = m^2c^4 + p^2c^2 \,, \tag{2.4}$$

where m is the particle rest mass and p the particle momentum. If the mass–energy dominates, the particle will be moving at much less than the speed of light, and we say it is non-relativistic. In that limit we can carry out an expansion

$$E_{\rm total} = mc^2 \left(1 + \frac{p^2}{m^2c^2}\right)^{1/2} \approx mc^2 + \frac{1}{2}\frac{p^2}{m} \,. \tag{2.5}$$

We recognize the first term as Einstein's famous $E = mc^2$, known as the rest mass–energy as it is the energy of the particle when it is stationary. The second term is the usual kinetic energy ($p = mv$ in the non-relativistic limit). If the mass–energy does not dominate, the particle will be moving at a substantial fraction of the speed of light and so is relativistic. In particular, any particle with zero rest mass is always relativistic and moves at the speed of light, the simplest example being light itself.

Let's review the nature of the particles which are believed to exist in our Universe.

Baryons

We ourselves are built from atoms, the bulk of whose mass is attributable to the protons and neutrons in the atomic nuclei. Protons and neutrons are believed to be made up of more fundamental particles known as quarks, a proton being made of two up quarks and a down quark, while a neutron is an up and two downs. A general term for particles made up of three quarks is **baryons**. Of all the possible baryons, only the proton and neutron can be stable,[3] and so these are thought to be the only types of baryonic particle significantly represented in the Universe. Yet another piece of terminology, nucleon, refers to just protons and neutrons, but I'll follow the standard practice of using the term baryon. In particle physics units, the mass–energies of a proton and a neutron are 938.3 MeV and 939.6 MeV respectively, where 'MeV' is a Mega-electron volt, a unit of energy equal to a million electron volts (eV) and rather more convenient than a Joule.

Although electrons are certainly not made from quarks, they are traditionally also included under the title baryon by cosmologists (to the annoyance of particle physicists). A crucial property of the Universe is that it is charge neutral, so there must be one electron for every proton. Weighing in at a puny 0.511 MeV, well under a thousandth of a proton mass, the contribution of electrons to the total mass is a tiny fraction, not meriting separate discussion.

In the present Universe, baryons are typically moving non-relativistically, meaning that their kinetic energy is much less than their mass–energy.

Radiation

Our visual perception of the Universe comes from electromagnetic radiation, and such radiation, at a large variety of frequencies, pervades the Universe. In the quantum mechanical view of light, it can be thought of as made up of individual particles — like packets of energy — known as **photons** and usually indicated by the symbol γ. Photons propagate, naturally enough, at the speed of light; since they have zero rest mass their total energy is always given by their kinetic energy, and is related to their frequency f by

$$E = hf \,, \tag{2.6}$$

where h is Planck's constant.

Photons can interact with the baryons and electrons; for example, a high-energy photon can knock an electron out of an atom (a process known as ionization), or can scatter off a free electron (known as Thomson scattering in the non-relativistic case $hf \ll m_{\mathrm{e}}c^2$, otherwise Compton scattering). The more energetic the photons are, the more devastating their effects on other particles.

---

[3]The proton lifetime is known to be either infinite, corresponding to the proton being absolutely stable, or much longer than the age of the Universe so that they are effectively stable. Isolated neutrons are unstable (decaying into a proton, an electron and an anti-neutrino), but those bound in nuclei may be stable; this will prove crucial in Chapter 12.

Neutrinos

Neutrinos are extremely weakly interacting particles, produced for example in radioactive decay. There is now significant experimental evidence that they possess a non-zero rest mass, but it is unclear whether this mass might be large enough to have cosmological effects, and it remains a working assumption in cosmology to treat them as effectively massless. I will adopt that assumption for the main body of this book, and in that case they, like photons, are always relativistic. The combination of photons and neutrinos makes up the relativistic material in our Universe. Confusingly, sometimes the term 'radiation' is used to refer to all the relativistic material.

There are three types of neutrino, the electron neutrino, muon neutrino and tau neutrino, and if they are indeed all massless they should all exist in our Universe. Unfortunately, their interactions are so weak that for now there is no hope of detecting cosmological neutrinos directly. Originally their presence was inferred on purely theoretical grounds, though we will see that the existence of the cosmic neutrino background may be inferred indirectly by some cosmological observations.

Because they are so weakly interacting, the experimental limits on the neutrino masses, especially of the latter two types, are quite weak, and it is in fact perfectly possible that they are massive enough to be non-relativistic. The possible effects of neutrino masses are explored in Advanced Topic 3.

Dark matter

In this book we'll encounter one further kind of particle that may exist in our Universe, which is not part of the Standard Model of particle theory. It is known as dark matter, and its properties are highly uncertain and a matter of constant debate amongst cosmologists. We'll return to it in Chapter 9.

2.5.2 Thermal distributions and the black-body spectrum

I end this section with some discussion of the physics of radiation. If this is unfamiliar to you, the details aren't all that crucial, though some of the results will be used later in the book.

If particles are frequently interacting with one another, then the distribution of their energies can be described by equilibrium thermodynamics. In a thermal distribution, interactions are frequent, but a balance has been reached so that all interactions proceed equally frequently in both the forward and backward directions, so that the overall distribution of particle numbers and energies remains fixed. The number of particles of a given energy then depends only on the temperature.

The precise distribution depends on whether the particles considered are fermions, which obey the Pauli exclusion principle, or bosons, which do not. In this book the most interesting case is that of photons, which are bosons, and their characteristic distribution at temperature T is the **Planck** or **black-body** spectrum. Photons have two possible polarizations, and each has an occupation number per mode $\mathcal{N}$ given by the Planck function

$$\mathcal{N} = \frac{1}{\exp\left(hf/k_{\mathrm{B}}T\right) - 1}, \tag{2.7}$$

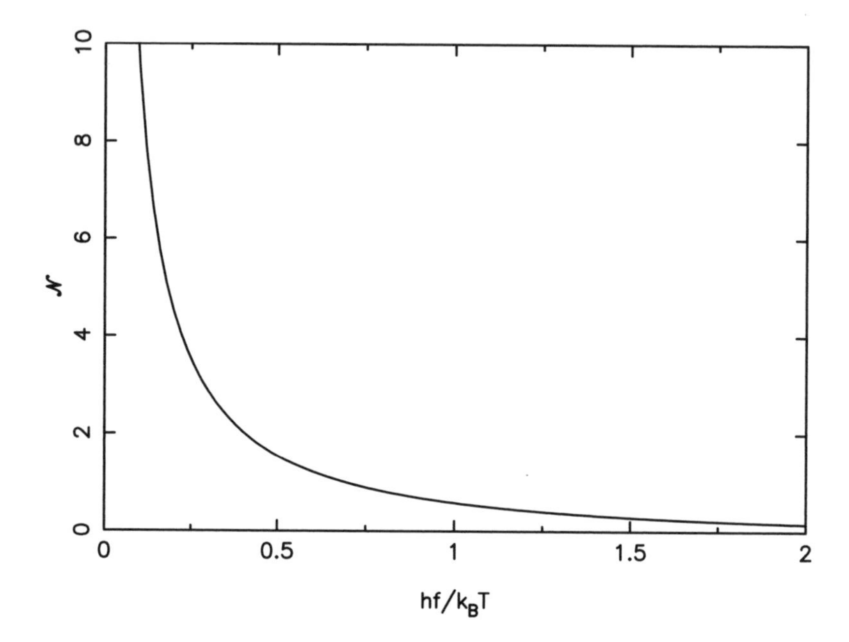

Figure 2.7 The Planck function of equation (2.7). There are far more photons with very low energy than very high energy.

where h is Planck's constant and $k_{\rm B}$ is one of the fundamental constants of Nature, the **Boltzmann constant**, whose value is $1.381 \times 10^{-23}\,{\rm J\,K^{-1}} = 8.619 \times 10^{-5}\,{\rm eV\,K^{-1}}$.

To interpret this equation, remember that hf is the photon energy. The purpose of the Boltzmann constant is to convert temperature into a characteristic energy $k_{\rm B}T$. Below this characteristic energy, $hf \ll k_{\rm B}T$, it is easy to make photons and the occupation number is large (as photons are bosons, the Pauli exclusion principle doesn't apply and there may be arbitrarily many photons in a given mode). Above the characteristic energy, $hf \gg k_{\rm B}T$, it is energetically unfavourable to make photons and the number is exponentially suppressed, as shown in Figure 2.7.

More interesting than the number of photons in a mode is the distribution of energy amongst the modes. We focus on the energy per unit volume, known as the **energy density** ϵ. Because there are very few photons with $hf \gg k_{\rm B}T$ there isn't much energy at high frequencies. But, despite their large number, there also isn't much total energy at low frequencies $hf \ll k_{\rm B}T$, both because those photons have less energy each ($E = hf$), and because their wavelength is longer and so each photon occupies a greater volume. With a considerable amount of work, the energy density in a frequency interval df about f can be shown to be

$$\epsilon(f)\,df = \frac{8\pi h}{c^3} \frac{f^3\,df}{\exp\left(hf/k_{\rm B}T\right) - 1}, \tag{2.8}$$

which tells us how the energy is distributed amongst the different frequencies. We see in Figure 2.8 that the peak of the distribution is at $f_{\rm peak} \simeq 2.8\,k_{\rm B}T/h$, corresponding to an energy $E_{\rm peak} = hf_{\rm peak} \simeq 2.8\,k_{\rm B}T$. That is to say, the total energy in the radiation is

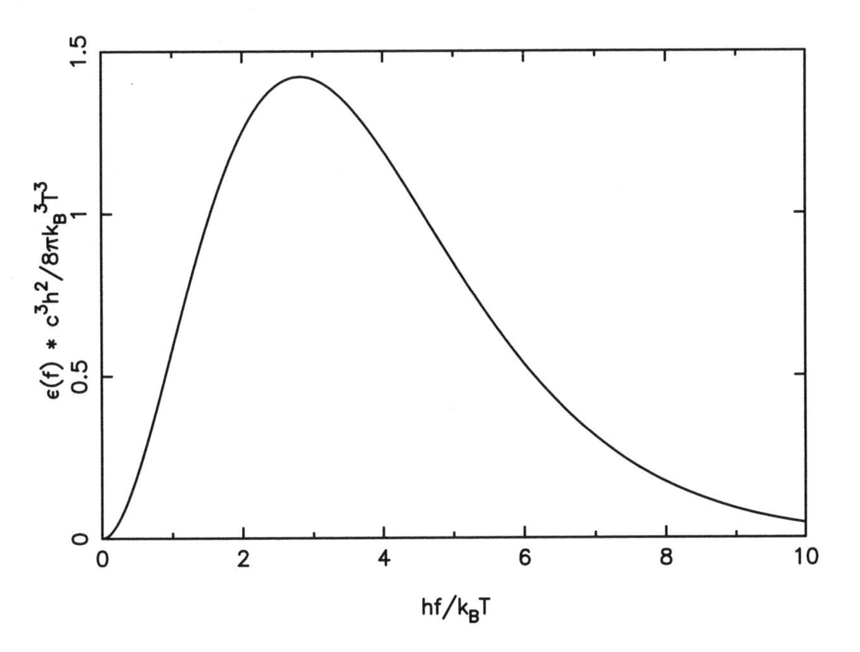

Figure 2.8 The energy density distribution of a black-body spectrum, given by equation (2.8). Most of the energy is contributed by photons of energy $hf \sim k_{\mathrm{B}}T$.

dominated by photons with energies of order $k_{\mathrm{B}}T$. Indeed, the mean energy of a photon in this distribution is $E_{\mathrm{mean}} \simeq 3\,k_{\mathrm{B}}T$.

When we study the early history of the Universe, an important question will be how this typical energy compares to atomic and nuclear binding energies.

A further quantity of interest will be the total energy density of the black-body radiation, obtained by integrating equation (2.8) over all frequencies. Setting $y = hf/k_{\mathrm{B}}T$ quickly leads to

$$\epsilon_{\mathrm{rad}} = \frac{8\pi k_{\mathrm{B}}^4}{h^3c^3}\,T^4 \times \int_0^\infty \frac{y^3\,dy}{e^y - 1}\,. \tag{2.9}$$

The integral is not particularly easy to compute, but you might like to try it as a challenge. The answer is $\pi^4/15$, giving a radiation energy density

$$\epsilon_{\mathrm{rad}} = \alpha T^4\,, \tag{2.10}$$

where the radiation constant α is defined as

$$\alpha = \frac{\pi^2 k_{\mathrm{B}}^4}{15\hbar^3 c^3} = 7.565 \times 10^{-16}\,\mathrm{J\,m^{-3}\,K^{-4}}\,. \tag{2.11}$$

Here $\hbar = h/2\pi$ is the reduced Planck constant.

Problems

2.1. Suppose that the Milky Way galaxy is a typical size, containing say 10^{11} stars, and that galaxies are typically separated by a distance of one megaparsec. Estimate the density of the Universe in SI units. How does this compare with the density of the Earth?

$1\,M_\odot \simeq 2 \times 10^{30}\,\text{kg}, 1\,\text{parsec} \simeq 3 \times 10^{16}\,\text{m}.$

2.2. In the real Universe the expansion is not completely uniform. Rather, galaxies exhibit some random motion relative to the overall Hubble expansion, known as their *peculiar velocity* and caused by the gravitational pull of their near neighbours. Supposing that a typical (e.g. root mean square) galaxy peculiar velocity is $600\,\text{km}\,\text{s}^{-1}$, how far away would a galaxy have to be before it could be used to determine the Hubble constant to ten per cent accuracy, supposing

(a) The true value of the Hubble constant is $100\,\text{km}\,\text{s}^{-1}\,\text{Mpc}^{-1}$?

(b) The true value of the Hubble constant is $50\,\text{km}\,\text{s}^{-1}\,\text{Mpc}^{-1}$?

Assume in your calculation that the galaxy distance and redshift could be measured exactly. Unfortunately, that is not true of real observations.

2.3. What evidence can you think of to support the assertion that the Universe is charge neutral, and hence contains an equal number of protons and electrons?

2.4. The binding energy of the electron in a hydrogen atom is 13.6 eV. What is the frequency of a photon with this energy? At what temperature does the mean photon energy equal this energy?

2.5. The peak of the energy density distribution of a black-body at $f_{\text{peak}} \simeq 2.8 k_{\text{B}} T/h$ implies that f_{peak}/T is a constant. Evaluate this constant in SI units (see page xiv for useful numbers). The Sun radiates approximately as a black-body with $T_{\text{sun}} \simeq 5800\,\text{K}$. Compute f_{peak} for solar radiation. Where in the electromagnetic spectrum does the peak emission lie?

2.6. The cosmic microwave background has a black-body spectrum at a temperature of $2.725\,\text{K}$. Repeat Problem 2.5 to find the peak frequency of its emission, and also find the corresponding wavelength and compare to Figure 2.4. Confirm that the peak emission lies in the microwave part of the electromagnetic spectrum. Finally, compute the total energy density of the microwave background.

Chapter 3

Newtonian Gravity

It is perfectly possible to discuss cosmology without having already learned general relativity. In fact, the most crucial equation, the Friedmann equation which describes the expansion of the Universe, turns out to be the same when derived from Newton's theory of gravity as it is when derived from the equations of general relativity. The Newtonian derivation is, however, some way from being completely rigorous, and general relativity is required to fully patch it up, a detail that need not concern us at this stage.

In Newtonian gravity all matter attracts, with the force exerted by an object of mass M on one of mass m given by the famous relationship

$$F = \frac{GMm}{r^2}, \tag{3.1}$$

where r is the distance between the objects and G is Newton's gravitational constant. That is, gravity obeys an inverse square law. Because a force on an object induces an acceleration which is also proportional to its mass, via $F = ma$, the acceleration an object feels under gravity is *independent* of its mass.

The force exerted means there is a gravitational potential energy

$$V = -\frac{GMm}{r}, \tag{3.2}$$

with the force exerted being in the direction which decreases the potential energy the fastest. Like the electric potential of two opposite charges, the gravitational potential is negative, favouring the two objects being close together. But with gravity there is no analogue of the repulsion of like charges. Gravity always attracts.

The derivation of the Friedmann equation requires a famous result due originally to Newton, which I won't attempt to prove here. This result states that in a spherically-symmetric distribution of matter, a particle feels no force at all from the material at greater radii, and the material at smaller radii gives exactly the force which one would get if all the material was concentrated at the central point. This property arises from the inverse square law; the same results exist for electromagnetism. One example of its use is that the gravitational (or electromagnetic) force outside a spherical object of unknown density profile depends only on the total mass (charge). Another is that an 'astronaut' inside a

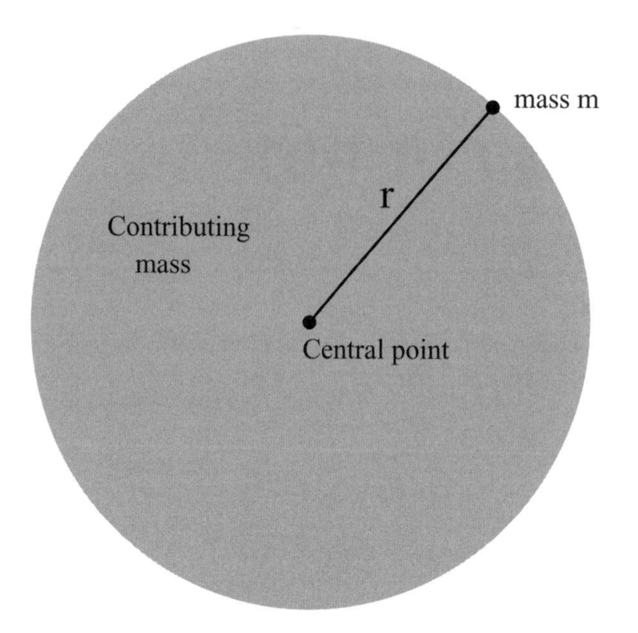

Figure 3.1 The particle at radius r only feels gravitational attraction from the shaded region. Any gravitational attraction from the material outside cancels out, according to Newton's theorem.

spherical shell feels no gravitational force, not only if they are at the centre but if they are at any position inside the shell.

3.1 The Friedmann equation

The Friedmann equation describes the expansion of the Universe, and is therefore the most important equation in cosmology. One of the routine tasks for a working cosmologist is solving this equation under different assumptions concerning the material content of the Universe. To derive the Friedmann equation, we need to compute the gravitational potential energy and the kinetic energy of a test particle (it doesn't matter which one, since everywhere in the Universe is the same according to the cosmological principle), and then use energy conservation.

Let's consider an observer in a uniform expanding medium, with mass density ρ, the mass density being the mass per unit volume. We begin by realizing that because the Universe looks the same from anywhere, we can consider any point to be its centre. Now consider a particle a distance r away with mass m, as shown in Figure 3.1. [By 'particle', I really mean a small volume containing the mass m.] Due to Newton's theorem, this particle only feels a force from the material at smaller radii, shown as the shaded region.

This material has total mass given by $M = 4\pi\rho r^3/3$, contributing a force

$$F = \frac{GMm}{r^2} = \frac{4\pi G\rho\, r m}{3}\,, \tag{3.3}$$

and our particle has a gravitational potential energy

$$V = -\frac{GMm}{r} = -\frac{4\pi G\rho\, r^2 m}{3}\,. \tag{3.4}$$

The kinetic energy is easy; the velocity of the particle is $\dot{r}$ (I'll always use dots to mean time derivatives) giving

$$T = \frac{1}{2}m\dot{r}^2\,. \tag{3.5}$$

The equation describing how the separation r changes can now be derived from energy conservation for that particle, namely

$$U = T + V\,, \tag{3.6}$$

where U is a constant. Note that U need not be the same constant for particles separated by different distances. Substituting gives

$$U = \frac{1}{2}\,m\dot{r}^2 - \frac{4\pi}{3}\,G\rho\, r^2 m\,. \tag{3.7}$$

This equation gives the evolution of the separation r between the two particles.

We now make a crucial step in this derivation, which is to realize that this argument applies to any two particles, because the Universe is homogeneous. This allows us to change to a different coordinate system, known as **comoving coordinates**. These are coordinates which are carried along with the expansion. Because the expansion is uniform, the relationship between real distance $\vec{r}$ and the comoving distance, which we can call $\vec{x}$, can be written

$$\vec{r} = a(t)\,\vec{x}\,, \tag{3.8}$$

where the homogeneity property has been used to ensure that a is a function of time alone. Note that these distances have been written as vector distances. What you should think of when studying this equation is a coordinate grid which expands with time, as shown in Figure 3.2. The galaxies remain at fixed locations in the $\vec{x}$ coordinate system. The original $\vec{r}$ coordinate system, which does not expand, is usually known as **physical coordinates**.

The quantity $a(t)$ is a crucial one, and is known as the **scale factor of the Universe**. It measures the universal expansion rate. It is a function of time alone, and it tells us how physical separations are growing with time, since the coordinate distances $\vec{x}$ are by definition fixed. For example, if, between times t_1 and t_2, the scale factor doubles in value, that tells us that the Universe has expanded in size by a factor two, and it will take us twice as long to get from one galaxy to another.

We can use the scale factor to rewrite equation (3.7) for the expansion. Substituting

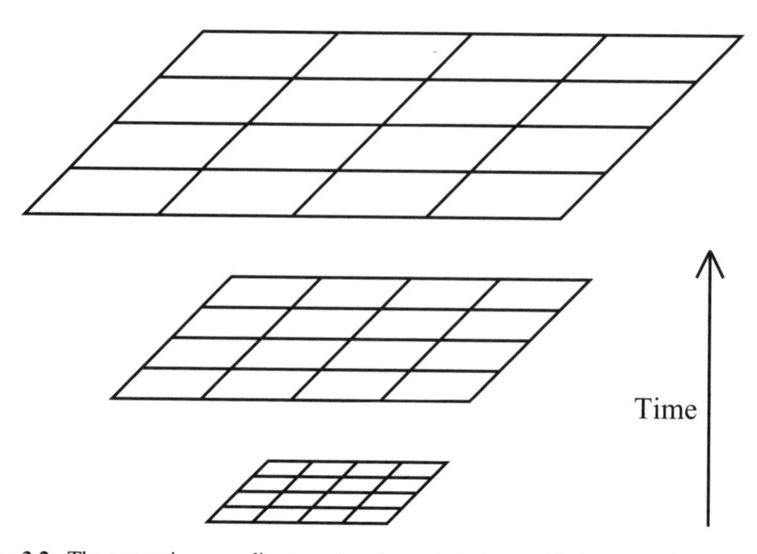

Figure 3.2 The comoving coordinate system is carried along with the expansion, so that any objects remain at fixed coordinate values.

equation (3.8) into it, remembering $\dot{x} = 0$ by definition as objects are fixed in comoving coordinates, gives

$$U = \frac{1}{2}m\dot{a}^2x^2 - \frac{4\pi}{3}G\rho\, a^2x^2m\,. \tag{3.9}$$

Multiplying each side by $2/ma^2x^2$ and rearranging the terms then gives

$$\left(\frac{\dot{a}}{a}\right)^2 = \frac{8\pi G}{3}\rho - \frac{kc^2}{a^2}\,, \tag{3.10}$$

where $kc^2 = -2U/mx^2$. This is the standard form of the **Friedmann equation**, and it will appear frequently throughout this book. In this expression k must be independent of x since all the other terms in the equation are, otherwise homogeneity will not be maintained. So in fact we learn that homogeneity requires that the quantity U, while constant for a given particle, does indeed change if we look at different separations x, with $U \propto x^2$.

Finally, since $k = -2U/mc^2x^2$ which is time independent (as the total energy U is conserved, and the comoving separation x is fixed), we learn that k is just a constant, unchanging with either space or time. It has the units of [length]$^{-2}$. An expanding Universe has a unique value of k, which it retains throughout its evolution. In Chapter 4 we will see that k tells us about the geometry of the Universe, and it is often called the curvature.

3.2 On the meaning of the expansion

So what does the expansion of the Universe mean? Well, let's start with what it does *not* mean. It does not mean that your body is gradually going to get bigger with time (and certainly isn't an excuse if it does). It does not mean that the Earth's orbit is going to get further from the Sun. It doesn't even mean that the stars within our galaxy are going to become more widely separated with time.

But it does mean that distant galaxies are getting further apart.

The distinction is whether or not the motion of objects is governed by the cumulative gravitational effect of a homogeneous distribution of matter between them, as shown in Figure 3.1. The atoms in your body are not; their separation is dictated by the strength of chemical bonds, with gravity playing no significant role. So molecular structures will not be affected by the expansion. Likewise, the Earth's motion in its orbit is completely dominated by the gravitational attraction of the Sun (with a minor contribution from the other planets). And even the stars in our galaxy are orbiting in the common gravitational potential well which they themselves create, and are not moving apart relative to one another. The common feature of these environments is that they are ones of considerable excess density, very different from the smooth distribution of matter we used to derive the Friedmann equation.

But if we go to large enough scales, in practice tens of megaparsecs, the Universe does become effectively homogeneous and isotropic, with the galaxies flying apart from one another in accordance with the Friedmann equation. It is on these large scales that the expansion of the Universe is felt, and on which the cosmological principle applies.

3.3 Things that go faster than light

A common question that concerns people is whether faraway galaxies are receding from us faster than the speed of light. That is to say, if velocity is proportional to distance, then if we consider galaxies far enough away can we not make the velocity as large as we like, in violation of special relativity?

The answer is that indeed in our theoretical predictions distant objects can *appear* to move away faster than the speed of light. However, it is space itself which is expanding. There is no violation of causality, because no signal can be sent between such galaxies. Further, special relativity is not violated, because it refers to the relative speeds of objects passing each other, and cannot be used to compare the relative speeds of distant objects.

One way to think of this is to imagine a colony of ants on a balloon. Suppose that the fastest the ants can move is a centimetre per second. If any two ants happen to pass each other, their fastest relative speed would be two centimetres per second, if they happened to be moving in opposite directions. Now start to blow the balloon up. Although the ants wandering around the surface still cannot exceed one centimetre per second, the balloon is now expanding under them, and ants which are far apart on the balloon could easily be moving apart at faster than two centimetres per second if the balloon is blown up fast enough. But if they are, they will never get to tell each other about it, because the balloon is pulling them apart faster than they can move together, even at full speed. Any ants that start close enough to be able to pass one another must do so at no more than two centimetres per second even if the Universe is expanding.

The expansion of space is just like that of the balloon, and pulls the galaxies along with it.

3.4 The fluid equation

Fundamental though it is, the Friedmann equation is of no use without an equation to describe how the density ρ of material in the Universe is evolving with time. This involves the pressure p of the material, and is called the fluid equation. [Unfortunately the standard symbol p for pressure is the same as for momentum, which we've already used. Almost always in this book, p will be pressure.] As we'll shortly see, the different types of material which might exist in our Universe have different pressures, and lead to different evolution of the density ρ.

We can derive the fluid equation by considering the first law of thermodynamics

$$dE + pdV = TdS\,, \tag{3.11}$$

applied to an expanding volume V of unit comoving radius.[1] This is exactly the same as applying thermodynamics to a gas in a cylinder. The volume has physical radius a, so the energy is given, using $E = mc^2$, by

$$E = \frac{4\pi}{3}a^3\rho\, c^2\,. \tag{3.12}$$

The change of energy in a time dt, using the product rule, is

$$\frac{dE}{dt} = 4\pi a^2 \rho\, c^2\,\frac{da}{dt} + \frac{4\pi}{3}a^3\,\frac{d\rho}{dt}\,c^2\,, \tag{3.13}$$

while the rate of change in volume is

$$\frac{dV}{dt} = 4\pi a^2\,\frac{da}{dt}\,. \tag{3.14}$$

Assuming a reversible adiabatic expansion $dS = 0$, putting these into equation (3.11) and rearranging gives

$$\dot{\rho} + 3\frac{\dot{a}}{a}\left(\rho + \frac{p}{c^2}\right) = 0\,, \tag{3.15}$$

where as always dots are shorthand for time derivatives. This is the **fluid equation**. As we see, there are two terms contributing to the change in the density. The first term in the brackets corresponds to the dilution in the density because the volume has increased, while the second corresponds to the loss of energy because the pressure of the material has done work as the Universe's volume increased. This energy has not disappeared entirely of course; energy is always conserved. The energy lost from the fluid via the work done has gone into gravitational potential energy.

---

[1] Don't confuse V for volume with V for gravitational potential energy.

Let me stress that there are no pressure forces in a homogeneous Universe, because the density and pressure are everywhere the same. A pressure *gradient* is required to supply a force. So pressure does not contribute a force helping the expansion along; its effect is solely through the work done as the Universe expands.

We are still not in a position to solve the equations, because now we only know what ρ is doing if we know what the pressure p is. It is in specifying the pressure that we are saying what kind of material our model Universe is filled with. The usual assumption in cosmology is that there is a unique pressure associated with each density, so that $p \equiv p(\rho)$. Such a relationship is known as the **equation of state**, and we'll see two different examples in Chapter 5. The simplest possibility is that there is no pressure at all, and that particular case is known as (non-relativistic) matter.

Once the equation of state is specified, the Friedmann and fluid equations are all we need to describe the evolution of the Universe. However, before discussing this evolution, I am going to spend some time exploring some general properties of the equations, as well as devoting Chapter 4 to consideration of the meaning of the constant k. If you prefer to immediately see how to solve these equations, feel free to jump straight away to Sections 5.3 to 5.5, and come back to the intervening material later. On the way, you might want to glance at Section 3.6 to find out why a factor of c^2 mysteriously vanishes from the Friedmann equation between here and there.

3.5 The acceleration equation

The Friedmann and fluid equations can be used to derive a third equation, not independent of the first two of course, which describes the acceleration of the scale factor. By differentiating equation (3.10) with respect to time we obtain

$$2\,\frac{\dot{a}}{a}\,\frac{a\ddot{a}-\dot{a}^2}{a^2} = \frac{8\pi G}{3}\dot{\rho} + 2\frac{kc^2\dot{a}}{a^3}\,. \tag{3.16}$$

Substituting in for $\dot{\rho}$ from equation (3.15) and cancelling the factor $2\dot{a}/a$ in each term gives

$$\frac{\ddot{a}}{a} - \left(\frac{\dot{a}}{a}\right)^2 = -4\pi G\left(\rho + \frac{p}{c^2}\right) + \frac{kc^2}{a^2}\,, \tag{3.17}$$

and finally, using equation (3.10) again, we arrive at an important equation known as the **acceleration equation**

$$\frac{\ddot{a}}{a} = -\frac{4\pi G}{3}\left(\rho + \frac{3p}{c^2}\right)\,. \tag{3.18}$$

Notice that if the material has any pressure, this *increases* the gravitational force, and so further decelerates the expansion. I remind you that there are no forces associated with pressure in an isotropic Universe, as there are no pressure gradients.

The acceleration equation does not feature the constant k which appears in the Friedmann equation; it cancelled out in the derivation.

3.6 On mass, energy and vanishing factors of c^2

You should be aware that cosmologists have a habit of using mass density ρ and energy density ϵ interchangeably. They are related via Einstein's most famous equation as $\epsilon = \rho c^2$, and if one chooses so-called 'natural units' in which c is set equal to one, the two become the same. For clarity, however, I will be careful to maintain the distinction. Note that the phrase 'mass density' is used in Einstein's sense — it includes the contributions to the mass from the energy of the various particles, as well as any rest mass they might have.

The habit of setting $c = 1$ means that the Friedmann equation is normally written without the c^2 in the final term, so that it reads

$$\left(\frac{\dot{a}}{a}\right)^2 = \frac{8\pi G}{3}\rho - \frac{k}{a^2} \,. \tag{3.19}$$

The constant k then appears to have units $[\text{time}]^{-2}$ — setting $c = 1$ makes time and length units interchangeable. Since the practice of omitting the c^2 in the Friedmann equation is widely adopted in other cosmology textbooks, I will drop it for the remainder of this book too. In practice, it is a rare situation indeed where one has to be careful about this.

Chapter 4

The Geometry of the Universe

We now consider the real meaning of the constant k which appears in the Friedmann equation

$$\left(\frac{\dot{a}}{a}\right)^2 = \frac{8\pi G}{3}\rho - \frac{k}{a^2}. \tag{4.1}$$

While the Newtonian derivation in the last chapter introduced this as a measure of the energy per particle, the true interpretation, apparent in the context of general relativity, is that it measures the curvature of space. General relativity tells us that gravity is due to the curvature of four-dimensional space-time, and a full analysis can be found in any general relativity textbook. Here I will be purely descriptive, and focus on the interpretation of k as measuring the curvature of the three spatial dimensions. Further details of general relativistic cosmology can be found in Advanced Topic 1.

We have demanded that our model Universes be both homogeneous and isotropic. The simplest type of geometry which can have this property is what is called a **flat geometry**, in which the normal rules of Euclidean geometry apply. However, it turns out that the assumption of isotropy is not enough to demand that as the only choice. Instead, there are three possible geometries for the Universe, and they correspond to k being zero, positive or negative.

4.1 Flat geometry

Euclidean geometry is based on a set of simple axioms (e.g. a straight line is the shortest distance between two points), plus one more complex axiom which says that parallel straight lines remain a fixed distance apart. These are the basis for the standard laws of geometry, and lead to the following conclusions:

- The angles of a triangle add up to 180°.
- The circumference of a circle of radius r is $2\pi r$.

Such a geometry might well apply to our own Universe. If that is the case, then the Universe must be infinite in extent, because if it came to a definite edge then that would clearly violate the principle that the Universe should look the same from all points.[1]

A Universe with this geometry is often called a **flat Universe**.

4.2 Spherical geometry

Euclid always hoped that the more artificial final axiom could be proven from the others. It wasn't until the 19th century that Riemann demonstrated that Euclid's final axiom was an arbitrary choice, and that one could make other assumptions. In doing so, he founded the subject of non-Euclidean geometry, which forms the mathematical foundation for Einstein's theory of general relativity.

The simplest kind of non-Euclidean geometry is actually very familiar to us; it is the spherical geometry which we use, for instance, to navigate around the Earth. Before worrying about the Universe having three dimensions, let's examine the properties of the two-dimensional surface of the Earth, shown in Figure 4.1.

First of all, we know that a perfect sphere looks the same from all points on its surface, so the condition of isotropy is satisfied (e.g. if someone hands you a snooker ball and asks which way up it is, you're not going to be able to tell them). But, unlike the case of a flat geometry, the spherical surface is perfectly finite in extent, its area being given by $4\pi r^2$. Yet there is no boundary, no 'edge' to the surface of the Earth. So it is perfectly possible to have a finite surface which nevertheless has no boundary.

If we draw parallel lines on the surface of the Earth, then they violate Euclid's final axiom. The definition of a straight line is the shortest distance between two points, which means that the straight lines in a spherical geometry are segments of great circles, such as the equator or the lines of longitude.[2] The lines of longitude are an excellent example of the failure of Euclid's axiom; as they cross the equator they are all parallel to one another, but rather than remaining a constant distance apart they meet at both poles.

If we draw a triangle on a sphere, we find that the angles do not add up to 180° degrees either. The easiest example to think about is to start at the North Pole. Draw two straight lines down to the equator, ninety degrees apart, and then join them with a line on the equator. You have drawn a triangle in which all three angles are 90°, shown in Figure 4.1.

The circumference of a circle also fails to obey the normal law. Suppose we draw a circle at a radius r from the North Pole, and we'll choose r so that our circle is the equator. That radius, measured on the surface of the sphere, corresponds to a quarter of a complete circle around the Earth, so $r = \pi R/2$ where R is the radius of the Earth. However, the circumference c is given by $2\pi R$, so instead of the usual relation one has $c = 4r$ for a circle drawn at the equator. The circumference is less than $2\pi r$. Problem 4.1 looks at the general case; you may find it helpful to glance at the figure on page 31 now.

Although I have only considered specific cases where the algebra is easy, it is true that whatever triangle or circle is drawn, you'll always find

- The angles of a triangle add up to more than 180°.
- The circumference of a circle is less than $2\pi r$.

---

[1] Well, that's almost true. See Advanced Topic 1.3 for a way to bypass that conclusion.

[2] Note that, apart from the equator, lines of latitude are *not* straight lines; this is why aeroplanes do not follow lines of latitude when flying, because they are not the shortest way to go!

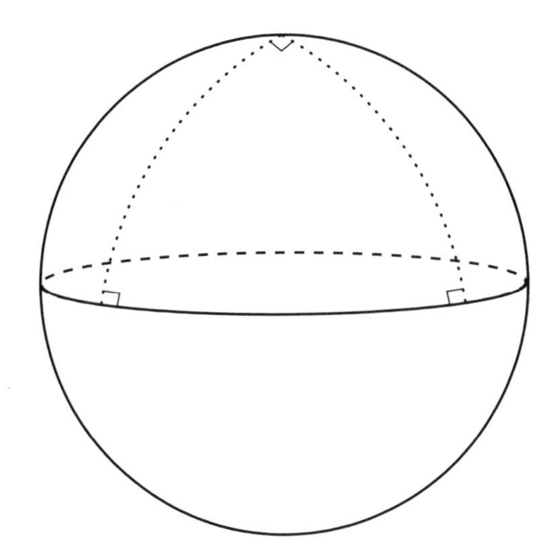

Figure 4.1 A sketch of a spherical surface, representing positive k. A triangle is shown which has three right angles!

If you make the triangles or circles much smaller than the size of the Earth, then the Euclidean laws start to become a good approximation; certainly we don't have to worry about Euclidean laws being broken in our everyday existence (though the appreciation that the Earth is spherical is vital for the planning of long distance journeys). So a small triangle drawn on a sphere will have the sum of its angles only marginally larger than 180°. This property makes it rather hard to measure the geometry of our own Universe, because the neighbouring region which we can measure accurately is only a small fraction of the size of the Universe and so will obey nearly Euclidean laws whatever the overall geometry.

One of the most important conceptual points that you need to grasp is that our three-dimensional Universe can have properties just like the two-dimensional surface of the sphere. Unfortunately our brains are not conditioned to think of three dimensions as being curved, so we must work by analogy with the two-dimensional situation I've just described. When we think of the surface of a sphere being curved, we naturally imagine the sphere as an object in our own three-dimensional Universe, and think of it as curved in that sense. The important point is that the curvature is a property of the two-dimensional surface of the sphere itself; the triangles and circles whose properties I've just described are drawn on the surface. A classic application of this was the ancient Greeks' use of these laws to deduce that the Earth is spherical, and even to obtain a good estimate of its diameter. When we discuss the spherical geometry, there is actually no need to think of it as existing within our three-dimensional space at all, and we must remember when discussing the geometry that we are assumed to be restricted to the surface of the sphere, and not allowed to move off the surface either towards or away from the notional centre.

All this is an analogy to what might happen with our three-dimensional Universe. The analogue of the two-dimensional sphere is called a three-sphere. In the rather unlikely event of four-dimensional creatures existing, they would be able to visualize the curvature of three-dimensional space in just the same way we can visualize that of two-dimensional

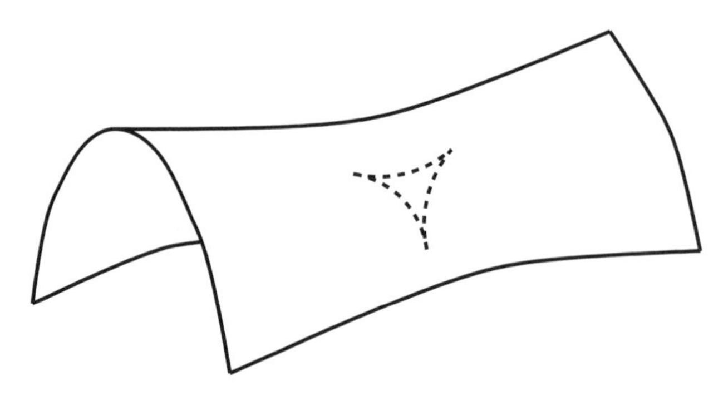

Figure 4.2 A sketch of a saddle surface, representing the hyperbolic geometry obtained when k is negative. A rather exaggerated triangle is shown with its sum of angles well below 180°.

space. However, like the two-dimensional sphere, the possible curvature of three-space is an intrinsic property and there is no actual need of a higher-dimensional space for it to live in. Obtaining a correct mental picture of this is one of the big challenges in understanding our Universe!

A Universe with a spherical geometry, like the surface of the Earth, has a finite size but no boundary. All points are equivalent. If we live in a spherical geometry, and travel in a straight line, we would not go on for ever and ever. Rather, eventually we would come back to where we had started from, from the opposite direction, exactly in the manner that someone travelling outward from the North Pole on the Earth eventually returns there from the opposite direction. Such a Universe corresponds to choosing a positive value for the quantity k appearing in the Friedmann equation. Because the special properties of the spherical geometry are due to its curvature, k is often therefore called the curvature term.

A Universe with $k > 0$ is normally referred to as a **closed Universe**, because of its finite size.

4.3 Hyperbolic geometry

The final choice is k negative. The corresponding geometry is known as hyperbolic, and is much less familiar than spherical geometry. It is normally represented by a saddle-like surface as in Figure 4.2; it is hard to see that this is consistent with isotropy but in fact it is. In a hyperbolic geometry, parallel lines never meet — in fact they break Euclid's axiom by diverging away from one another.

The behaviour of the hyperbolic geometry can be guessed from what has gone before; it is the opposite of the spherical geometry. One finds that

- The angles of a triangle add up to less than 180°.
- The circumference of a circle is greater than $2\pi r$.

Because parallel lines never meet, such a Universe must be infinite in extent, just as in the flat case. The situation $k < 0$ is known as an **open Universe**.

Table 4.1 A summary of possible geometries.

curvature	geometry	angles of triangle	circumference of circle	type of Universe
$k > 0$	spherical	$> 180^o$	$c < 2\pi r$	Closed
$k = 0$	flat	180°	$c = 2\pi r$	Flat
$k < 0$	hyperbolic	$< 180^o$	$c > 2\pi r$	Open

In Chapter 9 we'll explore which of the three possible geometries, summarized in Table 4.1, seems best suited to describe the real Universe.

4.4 Infinite and observable Universes

What does it actually mean for the Universe to be infinite, as in the flat and open cases? This property is nothing to do with the Universe lasting forever; what it means is that the Universe is already infinite in size even at a finite time. It genuinely goes on forever in all directions. Even despite this, it is still able to expand — the distances between objects can still increase regardless of whether the total Universe is infinite in extent or not. (You might consider the integers; these form an infinite set, but you can still multiply each by two to get a new infinite set where the separation between the numbers is twice as large.)

Nevertheless, this description is only a model, and we have no way of discovering whether the actual Universe does indeed go on forever. Cosmologists often talk about a different concept to the entire, possibly infinite, Universe, namely the **observable Universe**. This corresponds to the portion of the Universe we can actually see, and is limited by the finite speed of light. As the Universe gets older the observable Universe becomes larger and larger, by a combination of two effects. Firstly, the Universe is expanding, and secondly light has had longer to travel across the Universe. In practice, our knowledge of the Universe is restricted to this portion and we have no way of telling whether it does indeed continue into the infinite distance as required by the cosmological principle. It may be, for example, that the Universe becomes highly irregular on extremely large scales, and there are some theoretical models which predict that this might happen. Another possibility is discussed in Advanced Topic 1.3.

4.5 Where did the Big Bang happen?

A common question is 'where did the Big Bang happen?', suggesting perhaps that one could point in a specific direction and say 'That way!'. After all, in a conventional explosion that is a perfectly reasonable question to ask, as all the material flies outwards from the ignition point. Unfortunately, for the Big Bang things aren't so simple, and in a sense the answer is 'everywhere and nowhere'.

First of all, remember that our entire foundation is the cosmological principle, telling us that no point in the Universe is special. If there were a particular point where the 'Bang' happened, that would clearly be a special point and violate the cosmological principle. Rather, space and time themselves were created at the instant of the Big Bang (unlike a conventional explosion where the material flies through pre-existing space). If we take any point in the present Universe and trace back its history, it would start out at the explosion point, and in that sense the Big Bang happened everywhere in space.

In another sense, the location of the Big Bang is nowhere, because space itself is evolving and expanding, and it has changed since the Big Bang took place. Imagine the Universe as an expanding sphere; at any instant 'space' is the surface of the sphere, which is becoming bigger with time (again I'm thinking of a two-dimensional analogy to our real three-dimensional space). The place where the 'Bang' happened is at the centre of the sphere, but that's no longer part of the space, the surface of the sphere, in which we live. In particular, being constrained to the surface of the sphere means we are unable to 'point' to the place where the explosion is supposed to have happened. However, all the points in our current space were once at the centre of the sphere, when the Big Bang took place.

4.6 Three values of k

Because there are only three distinct possibilities for the geometry, many people indicate this explicitly by scaling their variables so that k takes on one of only three possible values, namely $k = -1$, 0 or $+1$, corresponding to the open, flat and closed cases respectively. This can be achieved by rescaling the scale factor by multiplying it by a fixed constant, namely $\hat{a} = a/\sqrt{|k|}$, in cases where k is non-zero. This leaves $H = \dot{a}/a$ unchanged, and removes the k from the final term in equation (4.1), which now reads

$$\left(\frac{\dot{\hat{a}}}{\hat{a}}\right)^2 = \frac{8\pi G}{3}\rho \pm \frac{1}{\hat{a}^2}, \qquad (4.2)$$

with '$-$' for positive k, '$+$' for negative k and the last term absent if $k = 0$.

If this rescaling is used, then (except in the $k = 0$ case) one loses the freedom to set $\hat{a} = 1$ at the present time, as will be used in Section 5.3. In effect, what one is doing is choosing to measure comoving distances in units of the so-called curvature scale, which is the scale on which the effects of curved space must be included, rather than in astronomical units such as megaparsecs.

In this book I will *not* rescale k to one of these three discrete values, though most of the time I will be specializing to $k = 0$ anyway.

Problems

4.1. Consider the surface of a two-dimensional sphere of radius R, as illustrated.

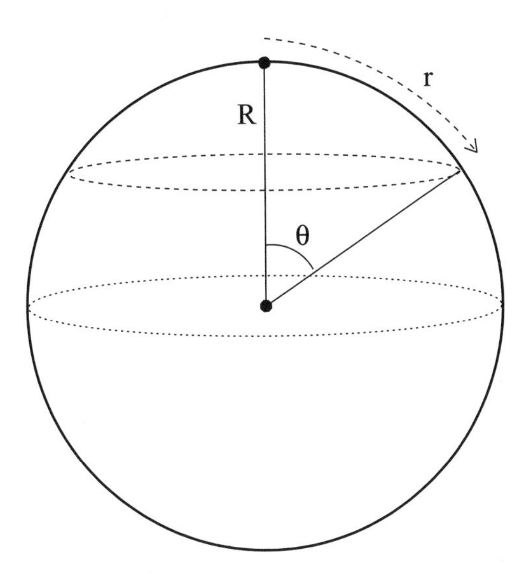

Circles are drawn on the sphere which have radius r *as measured on the surface on the sphere*, centred on the North Pole. Show that the general formula for the circumference c of such a circle, as a function of r, is given by

$$c = 2\pi \frac{\sin\theta}{\theta} r = 2\pi R \sin\frac{r}{R},$$

where θ is the angle between a line drawn from the centre of the sphere to the North Pole and one drawn to the circle. [Remember that by definition of angles in radians we have $r = \theta R$.]

Demonstrate that for small θ (i.e. $r \ll R$) this gives the normal flat geometry relation. Evaluate the relation for the case when the circle is at the equator.

4.2. Consider the spherical geometry of the previous problem, staying with the two-dimensional analogy to the real Universe. Suppose that galaxies are distributed evenly in such a Universe, with a number density n per unit area. Show that the total number N of galaxies inside a radius r is given by

$$N = 2\pi n R^2 \left[1 - \cos\frac{r}{R}\right].$$

Expand this for $r \ll R$ to show that the flat space result that the number is $n\pi r^2$ is recovered (remember we are working in only two dimensions). Do you see more or fewer galaxies out to the same radius, if the Universe is spherical rather than flat?

4.3. Is it possible for a closed Universe to evolve to become an open Universe? Give a reason for your answer.

Chapter 5

Simple Cosmological Models

In Chapter 3 we derived the equations satisfied by an expanding isotropic gas. They are the Friedman equation[1]

$$\left(\frac{\dot{a}}{a}\right)^2 = \frac{8\pi G}{3}\rho - \frac{k}{a^2}, \tag{5.1}$$

which governs the time evolution of the scale factor $a(t)$, and the fluid equation

$$\dot{\rho} + 3\frac{\dot{a}}{a}\left(\rho + \frac{p}{c^2}\right) = 0, \tag{5.2}$$

which gives us the evolution of the mass density $\rho(t)$. I stress that, despite their Newtonian derivation, these are the real equations used by cosmologists, more traditionally derived via the equations of general relativity as described in Advanced Topic 1.

This chapter finds and discusses some simple solutions to these equations, which will be used extensively during the book. However, while they have wide applicability during the early stages of the Universe, it will turn out that these are not sufficient to describe the present state of the Universe, for which an extra ingredient, the cosmological constant, will be needed. It is introduced in Chapter 7.

Before finding solutions to these equations, we can study two of their implications.

5.1 Hubble's law

The Friedmann equation allows us to explain Hubble's discovery that recession velocity is proportional to the distance. The velocity of recession is given by $\vec{v} = d\vec{r}/dt$ and is in the same direction as $\vec{r}$, allowing us to write

$$\vec{v} = \frac{|\dot{\vec{r}}|}{|\vec{r}|}\vec{r} = \frac{\dot{a}}{a}\vec{r}. \tag{5.3}$$

[1] In accord with the discussion of Section 3.6, the c^2 on the final term has been dropped, so that its appearance matches that of other cosmology textbooks.

The last step used $\vec{r} = a\vec{x}$, remembering that the comoving position $\vec{x}$ is a constant by definition. Consequently, Hubble's law $\vec{v} = H\vec{r}$ tells us that the proportionality constant, the Hubble parameter, should be identified as

$$H = \frac{\dot{a}}{a}, \tag{5.4}$$

and the value as measured today can be denoted with a subscript '0' as H_0. Because we measure Hubble's constant to be positive rather than negative, we know that the Universe is expanding rather than contracting.

We notice from this that the phrase Hubble's *constant* is a bit misleading. Although certainly it is constant in space due to the cosmological principle, there is no reason for it to be constant in time. In fact, using it as a more compact notation, we can write the Friedmann equation as an evolution equation for $H(t)$, as

$$H^2 = \frac{8\pi G}{3}\rho - \frac{k}{a^2}. \tag{5.5}$$

It is best to use the phrase 'Hubble parameter' for this quantity as a function of time, reserving 'Hubble constant' for its present value. Normally the Hubble parameter decreases with time, for instance as the expansion is slowed by the gravitational attraction of the matter in the Universe.

5.2 Expansion and redshift

The redshift of spectral lines that we used to justify the assumption of an expanding Universe can also be related to the scale factor. In this derivation I'll make the simplifying assumption that light is passed between two objects which are very close together, separated by a small distance dr, as shown in Figure 5.1. I've drawn the objects as galaxies, but I really mean two nearby points. According to Hubble's law, their relative velocity dv will be

$$dv = H\,dr = \frac{\dot{a}}{a}\,dr\,. \tag{5.6}$$

As the points are nearby we can directly apply the Doppler law to say that the change in wavelength between emission and reception, $d\lambda \equiv \lambda_{\rm r} - \lambda_{\rm e}$, is

$$\frac{d\lambda}{\lambda_{\rm e}} = \frac{dv}{c}, \tag{5.7}$$

where $d\lambda$ is going to be positive since the wavelength is increased. The time between emission and reception is given by the light travel time $dt = dr/c$, and putting all that together gives

$$\frac{d\lambda}{\lambda_{\rm e}} = \frac{\dot{a}}{a}\frac{dr}{c} = \frac{\dot{a}}{a}\,dt = \frac{da}{a}\,. \tag{5.8}$$

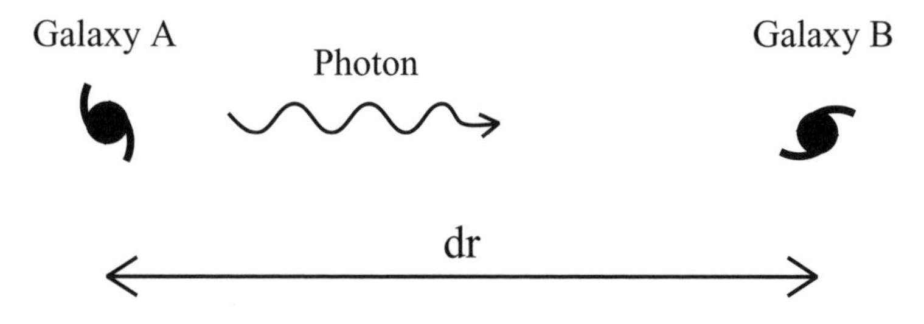

Figure 5.1 A photon travels a distance dr between two galaxies A and B.

Integrate and we find that $\ln \lambda = \ln a + \mathrm{constant}$, i.e.

$$\lambda \propto a\,, \tag{5.9}$$

where λ is now the instantaneous wavelength measured at any given time.

Although as I've derived it this result only applies to objects which are very close to each other, it turns out that it is completely general (a rigorous treatment is given in Advanced Topic 2). It tells us that as space expands, wavelengths become longer in direct proportion. One can think of the wavelength as being stretched by the expansion of the Universe, and its change therefore tells us how much the Universe has expanded since the light began its travels. For example, if the wavelength has doubled, the Universe must have been half its present size when the light was emitted.

The redshift as defined in equation (2.1) is related to the scale factor by

$$1 + z = \frac{\lambda_{\mathrm{r}}}{\lambda_{\mathrm{e}}} = \frac{a(t_{\mathrm{r}})}{a(t_{\mathrm{e}})}\,. \tag{5.10}$$

and is normally only used to refer to light received by us at the present epoch.

5.3 Solving the equations

In order to discover how the Universe might evolve, we need some idea of what is in it. In a cosmological context, this is done by specifying the relationship between the mass density ρ and the pressure p. This relationship is known as the **equation of state**. At this point, we shall only consider two possibilities.

Matter: In this context, the term 'matter' is used by cosmologists as shorthand for 'non-relativistic matter', and refers to any type of material which exerts negligible pressure, $p = 0$. Occasionally care is needed to avoid confusion between 'matter' used in this sense, and used to indicate all types of matter whether non-relativistic or not. A pressureless Universe is the simplest assumption that can be made. It is a good approximation to use for the atoms in the Universe once it has cooled down, as they are quite well separated and seldom interact, and it is also a good description of a collection of galaxies in the Universe, as they have no interactions other than gravitational ones. Occasionally the term 'dust' is used instead of 'matter'.

Radiation: Particles of light move, naturally enough, at the speed of light. Their kinetic energy leads to a pressure force, the radiation pressure, which using the standard theory of radiation can be shown to be $p = \rho c^2/3$. Problem 5.2 gives a rather hand-waving derivation of this result. More generally, any particles moving at highly-relativistic speeds have this equation of state, neutrinos being an obvious example.

I will concentrate on the case where the constant k in the Friedmann equation is set equal to zero, corresponding to a flat geometry.

5.3.1 Matter

We start by solving the fluid equation, having set $p = 0$ for matter. One way to solve it is to notice a clever way of rewriting it, as follows

$$\dot{\rho} + 3\frac{\dot{a}}{a}\rho = 0 \quad \Longrightarrow \quad \frac{1}{a^3}\frac{d}{dt}\left(\rho a^3\right) = 0 \quad \Longrightarrow \quad \frac{d}{dt}\left(\rho a^3\right) = 0\,, \tag{5.11}$$

though one could also solve it more formally by noting that it is a separable equation. Integrating tells us that ρa^3 equals a constant, i.e.

$$\rho \propto \frac{1}{a^3}\,. \tag{5.12}$$

This is not a surprising result. It says that the density falls off in proportion to the volume of the Universe. It is very natural that if the volume of the Universe increases by a factor of say two, then the density of the matter must fall by the same factor. After all, material cannot come from nowhere, and there is no pressure to do any work.

The equations we are solving (with $k = 0$) have one very useful symmetry; their form is unchanged if we multiply the scale factor a by a constant, since only the combination $\dot{a}/a$ appears. This means that we are free to rescale $a(t)$ as we choose, and the normal convention is to choose $a = 1$ at the present time. With this choice physical and comoving coordinate systems coincide at the present, since $\vec{r} = a\,\vec{x}$. Throughout this book I will use the subscript '0' to indicate the present value of quantities. Denoting the present density by ρ_0 fixes the proportionality constant

$$\rho = \frac{\rho_0}{a^3}\,. \tag{5.13}$$

Having solved for the evolution of the density in terms of a, we must now find how a varies with time by using the Friedmann equation. Substituting in for ρ, and remembering we are assuming $k = 0$, gives

$$\dot{a}^2 = \frac{8\pi G\rho_0}{3}\frac{1}{a}\,. \tag{5.14}$$

Faced with an equation like this, one can use formal techniques to solve it (this equation is separable, allowing it to be integrated), or alternatively make an educated guess as to the solution and confirm it by substitution. In cosmology, a good educated guess is normally a power-law $a \propto t^q$. Substituting this in, the left-hand side has time dependence t^{2q-2} and the right-hand side t^{-q}. This can only be a solution if these match, which requires $q = 2/3$, and so the solution is $a \propto t^{2/3}$. As we have fixed $a = 1$ at the present time $t = t_0$, the full solution is therefore

$$a(t) = \left(\frac{t}{t_0}\right)^{2/3} \quad ; \quad \rho(t) = \frac{\rho_0}{a^3} = \frac{\rho_0 t_0^2}{t^2} \,. \tag{5.15}$$

In this solution, the Universe expands forever, but the rate of expansion $H(t)$ decreases with time

$$H \equiv \frac{\dot{a}}{a} = \frac{2}{3t} \,, \tag{5.16}$$

becoming infinitely slow as the Universe becomes infinitely old. Notice that despite the pull of gravity, the material in the Universe does not recollapse but rather expands forever.

This is one of the classic cosmological solutions, and will be much used throughout this book.

5.3.2 Radiation

Radiation obeys $p = \rho c^2/3$. Consequently the fluid equation is changed from the matter-dominated case, now reading

$$\dot{\rho} + 4\frac{\dot{a}}{a}\rho = 0 \,. \tag{5.17}$$

This is amenable to the same trick as before, with the a^3 replaced by a^4 in equation (5.11), giving

$$\rho \propto \frac{1}{a^4} \,. \tag{5.18}$$

Carrying out the same analysis we did in the matter-dominated case gives

$$a(t) = \left(\frac{t}{t_0}\right)^{1/2} \quad ; \quad \rho(t) = \frac{\rho_0}{a^4} = \frac{\rho_0 t_0^2}{t^2} \,. \tag{5.19}$$

This is the second classic cosmological solution.

Notice that the Universe expands more slowly if radiation dominated than if matter dominated, a consequence of the extra deceleration that the pressure supplies — see equation (3.18). So it is definitely wrong to think of pressure as somehow 'blowing' the Universe apart. However, in each case the density of material falls off as t^2.

We'd better examine the fall off of the radiation density with volume more carefully. It drops as the fourth power of the scale factor. Three of those powers we have already identified as the increase in volume, leading naturally to a drop in the density. The final power arises from a different effect, the stretching of the wavelength of the light. Since the stretching is proportional to a, and the energy of radiation proportional to its frequency via $E = hf$, this results in a further drop in energy by the remaining power of a. This lowering of energy is exactly the redshifting effect we use to measure distances.

The rate of decrease of the radiation density also has an explanation in terms of thermodynamics, which is macroscopic rather than microscopic. Since the Universe in this case has a pressure, when it expands there is work done which is given by $p\,dV$, in exactly the same way as work is done on a piston when the gas is allowed to expand and cool. This work done corresponds to the extra diminution of the radiation density by the final factor of a.

5.3.3 Mixtures

A more general situation is when one has a mixture of both matter and radiation. Then there are two separate fluid equations, one for each of the two components. The trick which allows us to write ρ as a function of a still works, so we still have

$$\rho_{\mathrm{mat}} \propto \frac{1}{a^3} \quad ; \quad \rho_{\mathrm{rad}} \propto \frac{1}{a^4} \,. \tag{5.20}$$

However, there is still only a single Friedmann equation (after all, there is only one Universe!), which now has

$$\rho = \rho_{\mathrm{mat}} + \rho_{\mathrm{rad}} \,. \tag{5.21}$$

This means that the scale factor will have a more complicated behaviour, and so to convert $\rho(a)$ into $\rho(t)$ is much harder. It is possible to obtain exact solutions for this situation, but they are very messy so I won't include them here. Instead, I'll consider the simpler situation where one or other of the densities is by far the larger.

In that case, we can say that the Friedmann equation is accurately solved by just including the dominant component. That is, we can use the expansion rates we have already found. For example, suppose radiation is much more important. Then one would have

$$a(t) \propto t^{1/2} \quad ; \quad \rho_{\mathrm{rad}} \propto \frac{1}{t^2} \quad ; \quad \rho_{\mathrm{mat}} \propto \frac{1}{a^3} \propto \frac{1}{t^{3/2}} \,. \tag{5.22}$$

Notice that the density in matter falls off more slowly than that in radiation. So the situation of radiation dominating cannot last forever; however small the matter component might be originally it will eventually come to dominate. We can say that domination of the Universe by radiation is an unstable situation.

In the opposite situation, where it is the matter which is dominant, we obtain the solution

$$a(t) \propto t^{2/3} \quad ; \quad \rho_{\mathrm{mat}} \propto \frac{1}{t^2} \quad ; \quad \rho_{\mathrm{rad}} \propto \frac{1}{a^4} \propto \frac{1}{t^{8/3}} \,. \tag{5.23}$$

Matter domination is a stable situation, the matter becoming increasingly dominant over the radiation as time goes by.

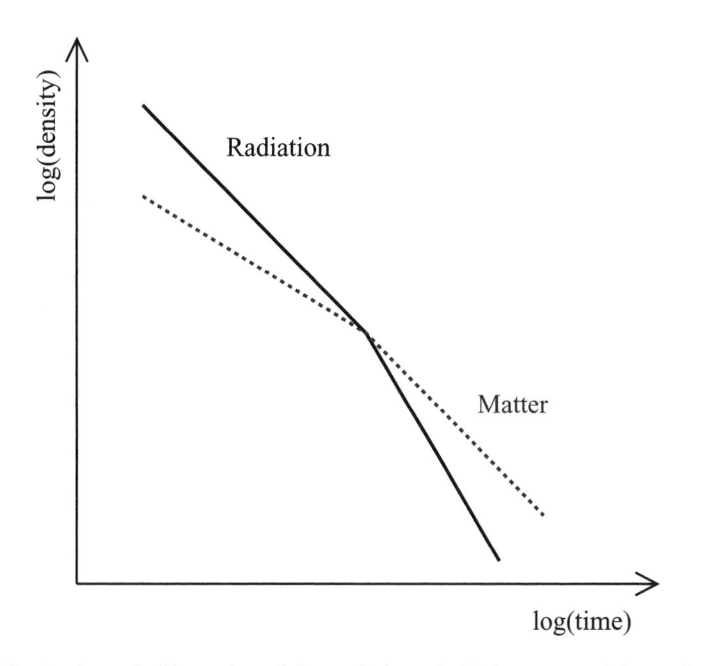

Figure 5.2 A schematic illustration of the evolution of a Universe containing radiation and matter. Once matter comes to dominate the expansion rate speeds up, so the densities fall more quickly with time.

Figure 5.2 shows the evolution of a Universe containing matter and radiation, with the radiation initially dominating. Eventually the matter comes to dominate, and as it does so the expansion rate speeds up from $a(t) \propto t^{1/2}$ to the $a(t) \propto t^{2/3}$ law. It is very possible that this is the situation which applies in our present Universe, as we'll see in Chapter 11.

5.4 Particle number densities

An important alternative view of the evolution of particles, which will be much used later in the book, is that of the **number density** n of particles rather than of their mass or energy density.

The number density is simply the number of particles in a given volume. If the mean energy per particle (including any mass–energy) is E, then the number density is related to the energy density by

$$\epsilon = n \times E \,. \tag{5.24}$$

The number density is useful because in most circumstances particle number is conserved. For example, if particle interactions are negligible, you wouldn't expect an electron to suddenly vanish into oblivion, and the same is true of a photon of light. The particle number can change through interactions, for example an electron and positron could annihilate and create two photons. However, if the interaction rate is high we expect the Universe to be in a state of thermal equilibrium. If so, then particle number is conserved even in a highly-interacting state, since by definition thermal equilibrium means that any interaction, which may change the number density of a particular type of particle, must proceed at the same rate in both forward and backward directions so that any change cancels out.

So, barring brief periods where thermal equilibrium does not hold, we expect the number of particles to be conserved. The only thing that changes the number *density*, therefore, is that the volume is getting bigger, so that these particles are spread out in a larger volume. This implies

$$n \propto \frac{1}{a^3}. \tag{5.25}$$

This looks encouragingly like the behaviour we have already seen for matter, but it's also true for radiation as well!

How does this relate to our earlier results? The energy of non-relativistic particles is dominated by their rest mass–energy which is constant, so

$$\rho_{\mathrm{mat}} \propto \epsilon_{\mathrm{mat}} \propto n_{\mathrm{mat}} \times E_{\mathrm{mat}} \propto \frac{1}{a^3} \times \mathrm{const} \propto \frac{1}{a^3}. \tag{5.26}$$

But photons lose energy as the Universe expands and their wavelength is stretched, so their energy is $E_{\mathrm{rad}} \propto 1/a$ as we have already seen. So

$$\rho_{\mathrm{rad}} \propto \epsilon_{\mathrm{rad}} \propto n_{\mathrm{rad}} \times E_{\mathrm{rad}} \propto \frac{1}{a^3} \times \frac{1}{a} \propto \frac{1}{a^4}. \tag{5.27}$$

These are exactly the results we saw before, equations (5.12) and (5.18).

Although the energy densities of matter and radiation evolve in different ways, their particle numbers evolve in the same way. So, apart from epochs during which the assumption of thermal equilibrium fails, the relative number densities of the different particles (e.g. electrons and photons) do not change as the Universe expands.

5.5 Evolution including curvature

We can now re-introduce the possibility that the constant k is non-zero, corresponding to spherical or hyperbolic geometry. Rather than seeking precise solutions, I will concentrate on the qualitative properties of the solutions. These are actually of rather limited use in describing our own Universe, because as we will see the cosmological models discussed so far are not general enough and we will need to consider a cosmological constant (see Chapter 7). Nevertheless, studying the possible behaviours of these simple models is a useful exercise, even if one should be cautious about drawing general conclusions.

In analyzing the possible dynamics, I will assume that the Universe is dominated by non-relativistic matter always, which in practice is not a restrictive assumption. We have already seen that if we assume that the constant k in the Friedmann equation is zero, then the Universe expands for ever, $a \propto t^{2/3}$, but slows down arbitrarily at late times. So we know the fate of the Universe in that case. But what happens if $k \neq 0$?

The principal question to ask is whether it is possible for the expansion of the Universe

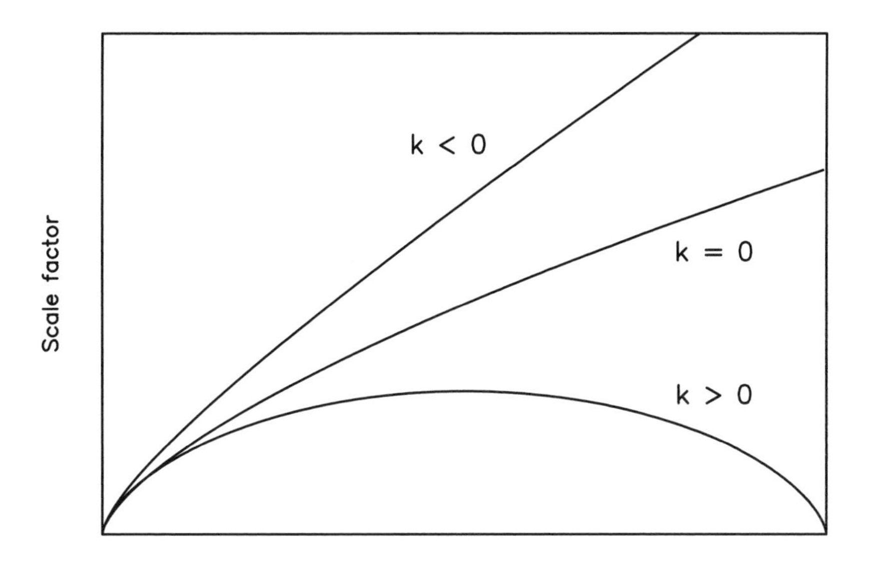

Figure 5.3 Three possible evolutions for the Universe, corresponding to the different signs of k. The middle line corresponds to the $k = 0$ case where the expansion rate approaches zero in the infinite future. During the early phases of the expansion the lines are very close and so observationally it can be difficult to distinguish which path the actual Universe will follow.

to stop, which since $H = \dot{a}/a$ corresponds to $H = 0$. Looking at the Friedmann equation

$$H^2 = \frac{8\pi G}{3}\rho - \frac{k}{a^2}, \tag{5.28}$$

it is immediately apparent that this is not possible if k is negative, for then both the terms on the right-hand side of the Friedmann equation are positive. Consequently, such a Universe must expand forever. That enables us to study the late-time behaviour, because we can see that the term k/a^2 falls off more slowly with the expansion than does $\rho_{\rm mat} \propto 1/a^3$. Since a becomes arbitrarily large for the matter-dominated solution for negligible k, the k/a^2 term must eventually come to dominate. When it does, the Friedmann equation becomes

$$\left(\frac{\dot{a}}{a}\right)^2 = -\frac{k}{a^2}. \tag{5.29}$$

Cancel off the a^2 terms and you'll find the solution is $a \propto t$. So when the last term comes to dominate, the expansion of the Universe becomes yet faster. In this case, the velocity does not tend to zero at late times, but instead becomes constant. This is sometimes known as free expansion.

Things are very different if k is positive. It then becomes possible for H to be zero, by the two terms on the right-hand side of the Friedmann equation cancelling each other out. Indeed, this is inevitable, because the negative influence of the k/a^2 term will become more and more important relative to the $\rho_{\rm mat}$ term as time goes by. In such a Universe therefore, the expansion must come to an end after a finite amount of time. As gravitational attraction persists, the recollapse of the Universe becomes inevitable.

In fact, the collapse of the Universe is fairly easy to describe, because the equations governing the evolution are time reversible. That is, if one substitutes $-t$ for t, they remain the same. The collapse phase is therefore just like the expansion in reverse, and so after a finite time the Universe will come to an end in a Big Crunch. Problem 5.5 investigates this in more detail.

These three behaviours, illustrated in Figure 5.3, can be related to the particle energy U in our Newtonian derivation of the Friedmann equation. If the particle energy is positive, then it can escape to infinity, with a final kinetic energy given by U. If the total energy is zero, then the particle can just escape but with zero velocity. Finally, if the energy is negative, it cannot escape the gravitational attraction and is destined to recollapse inwards.

There is a fairly precise analogy with escape velocity from the Earth (or the Moon, if you want to worry about the atmosphere). If you throw a rock up in the air hard enough, gravity will be unable to stop it and eventually it will sail off into space at a constant velocity. If your throw is too puny, it will rapidly fall back. And in between is the escape velocity, where the rock is just able to escape the gravitational field and no more.

Problems

5.1. Is the total energy of the Universe conserved as it expands?

5.2. This problem indicates the origin of the equation of state $p = \rho c^2/3$ for radiation. An ideal gas has pressure

$$p = \frac{1}{3} n \langle \mathbf{v}.\mathbf{p} \rangle \,,$$

where $\langle \cdots \rangle$ indicates an average over the direction of particle motions. Here n is the number density, and be careful not to confuse the unfortunate notation p for pressure and $\mathbf{p}$ for momentum. Using equation (2.4) to relate the photon energy and momentum, show that this gives

$$p = \frac{1}{3} n \langle E \rangle \,,$$

where $\langle E \rangle$ is the mean photon energy. Hence demonstrate the equation of state for radiation.

5.3. During this chapter we examined solutions for the expansion when the Universe contained either matter ($p = 0$) or radiation ($p = \rho c^2/3$). Suppose we have a more general equation of state, $p = (\gamma - 1)\rho c^2$, where γ is a constant in the range $0 < \gamma < 2$. Find solutions for $\rho(a)$, $a(t)$ and hence $\rho(t)$ for Universes containing such matter. Assume $k = 0$ in the Friedmann equation.
What is the solution if $p = -\rho c^2$?

5.4. Using your answer to Problem 5.3, what value of γ would be needed so that ρ has the same time dependence as the curvature term k/a^2? Find the solution $a(t)$ to the full Friedmann equation for a fluid with this γ, assuming negative k.

5.5. The full Friedmann equation is

$$\left(\frac{\dot{a}}{a}\right)^2 = \frac{8\pi G}{3}\rho - \frac{k}{a^2} .$$

Consider the case $k > 0$, with a Universe containing only matter ($p = 0$) so that $\rho = \rho_0/a^3$. Demonstrate that the parametric solution

$$a(\theta) = \frac{4\pi G\rho_0}{3k}(1 - \cos\theta) \quad ; \quad t(\theta) = \frac{4\pi G\rho_0}{3k^{3/2}}(\theta - \sin\theta)$$

solves this equation, where θ is a variable which runs from 0 to 2π.
Sketch a and t as functions of θ. Describe qualitatively the behaviour of the Universe. Attempt to sketch a as a function of t.

5.6. Now consider the case $k < 0$, with a Universe containing only matter ($p = 0$) so that $\rho = \rho_0/a^3$. What is the solution $a(t)$ in a situation where the final term of the Friedmann equation dominates over the density term? How does the density of matter vary with time? Is domination by the curvature term a stable situation that will continue forever?

Chapter 6

Observational Parameters

The Big Bang model does not give a unique description of our present Universe, but rather leaves quantities such as the present expansion rate, or the present composition of the Universe, to be fixed by observation.

It is a standard practice to specify cosmological models via a few parameters, which one then tries to determine observationally to decide which version of the model best describes our Universe. In this chapter and the next I'll discuss the most commonly-considered parameters, including ones we have already seen and new ones.

6.1 The expansion rate H_0

The Hubble constant H_0, which tells us the present expansion rate of the Universe, is in many ways the most fundamental cosmological parameter of all. It also ought to be the easiest to measure, since all galaxies are supposed to obey $v = H_0 r$. So all we have to do is measure the velocities and distances of as many galaxies as we can and get an answer. However, each measurement has its problems.

Velocities are given by the redshift of spectral lines, a measurement which is now easy enough that the velocity of an individual galaxy can be measured to high accuracy. However, remember that the cosmological principle isn't perfect, and so, as well as the uniform expansion we are trying to measure, galaxies also have motions relative to one another, the so-called peculiar velocity. The peculiar velocities are randomly oriented, and for a given galaxy we cannot split its measured velocity into the Hubble expansion and the peculiar velocity. However, the cosmological principle does tell us that the typical size of the peculiar velocity should not depend on where in the Universe the galaxy is. It is therefore independent of distance, whereas the Hubble velocity is proportional to distance. If we look far enough away (in practice many tens of megaparsecs) then the Hubble velocity dominates and the (unknown) peculiar velocity can be ignored.

Given that the expansion velocity can only be accurately distinguished from the peculiar velocity at large distances, we need to be able to estimate these large distances accurately in order to carry out the calculation $H_0 = v/r$. These distances are much harder to obtain, because galaxies are far too distant to be located by parallax. [Remember that an object one parsec away has a parallax (i.e. an apparent motion when viewed from different parts of Earth's orbit) of one arcsecond, by definition. A galaxy many megaparsecs away will have an immeasurably small parallax of less than a micro-arcsecond.] The usual method is known as 'standard candles', where some type of object is assumed to have exactly the same properties in all parts of the Universe. This is the cosmological equivalent of saying that if one light bulb looks a quarter as bright as another, then from the inverse square law it must be twice as far away — fine as long as you believe that all light bulbs have precisely the same brightness. A classic example is the period–luminosity relation in cepheid variable stars. The period of variability of those stars is readily measured, and there is reasonable empirical evidence of a relation between the period and the luminosity of the stars which lets us convert the measured period into a brightness. Other standard candles which have been used include the brightness of certain types of supernovae, and the brightest galaxies in galaxy clusters. All these methods have good success at determining relative distances between two galaxies, which requires that the objects be good standard candles, and relative distances are all that is needed to confirm the Hubble law.

However, to give an absolute distance, actually measuring the proportionality constant H_0, we also need a calibration against an object of known distance, which proves much harder. In the light bulb analogy, to get relative distances we need only the inverse square law and the belief that all bulbs have the same brightness; we don't need to know how bright the bulbs are. But to be able to say how far away a bulb of a given observed brightness is, we need to know its absolute brightness.

It is only very recently, principally through efforts of a team led by Wendy Freedman using the Hubble Space Telescope, that the calibration problem has begun to come under control. Even so, the Hubble constant is still not known as accurately as we would like, although the Hubble law of expansion is extremely well determined. The Hubble constant is usually parametrized as

$$H_0 = 100\,h\ \mathrm{km\,s^{-1}\,Mpc^{-1}}\,, \tag{6.1}$$

and the final result from the Hubble Space Telescope Key Project gives

$$h = 0.72 \pm 0.08\,, \tag{6.2}$$

where the uncertainty is a one-sigma error (meaning it should be doubled to indicate 95 percent confidence, at least if the uncertainty is approximately gaussian-distributed). If the value is indeed $h = 0.72$ then an object with a recession velocity of $7200\ \mathrm{km\,s^{-1}}$ would be expected to be at a distance $v/H_0 = 100$ Mpc. The smaller the value of h, the more slowly the Universe is expanding.

That h is not more accurately determined introduces uncertainties throughout cosmology. In particular, the actual distances to faraway objects are only known up to an uncertainty of the factor h, because recession velocities are the only way to estimate their distance. For this reason it is common to see distances specified in the form, for example, $100\,h^{-1}$ Mpc, where the number is accurately known but h^{-1} is not. You'll see factors of h cropping up frequently in the rest of this book.

This situation is analogous to being given a map without a scale. Suppose you happen to find yourself on Sauchiehall Street in Glasgow. A map without a scale is perfectly good at telling you that the coffee shop is twice as far away as the cinema, but won't tell you the distance to either. However, if you walk to the cinema and find that the distance is 146 metres, you then know the distance not only to the cinema, but also the coffee shop and for that matter the distance to the pub on the corner too, because you've calibrated your map. In cosmology, however, there's no good way to hike out to the Coma galaxy cluster to find its distance, so our knowledge of the scale of our maps of the Universe, such as the one shown in Figure 2.2, remains imprecise.

6.2 The density parameter Ω_0

The density parameter is a very useful way of specifying the density of the Universe. Let's start with the Friedmann equation again. Recalling that $H = \dot{a}/a$, it reads

$$H^2 = \frac{8\pi G}{3}\rho - \frac{k}{a^2}\,. \tag{6.3}$$

For a given value of H, there is a special value of the density which would be required in order to make the geometry of the Universe flat, $k = 0$. This is known as the **critical density** ρ_{c}, which we see is given by

$$\rho_{\mathrm{c}}(t) = \frac{3H^2}{8\pi G}\,. \tag{6.4}$$

Note that the critical density changes with time, since H does. Since we know the present value of the Hubble constant [at least in terms of h defined in equation (6.1)], we can compute the present critical density. Since $G = 6.67 \times 10^{-11}\,\mathrm{m}^3\,\mathrm{kg}^{-1}\,\mathrm{sec}^{-2}$, and converting megaparsecs to metres using conversion factors quoted on page xiv, it is

$$\rho_{\mathrm{c}}(t_0) = 1.88\,h^2 \times 10^{-26}\,\mathrm{kg\,m}^{-3}\,. \tag{6.5}$$

This is a startlingly small number; compare for example the density of water which is $10^3\,\mathrm{kg\,m}^{-3}$. If there is any more matter than this apparently tiny amount, it is enough to tip the balance beyond a flat Universe to a closed one with $k > 0$. So only a very tiny density of matter is needed in order to provide enough gravitational attraction to halt and reverse the expansion of the Universe.

However, let us write that another way, since kilograms and metres are rather small and inconvenient units for dealing with something as big as the Universe. Let's try measuring masses in solar masses and distances in megaparsecs. It becomes

$$\rho_{\mathrm{c}}(t_0) = 2.78\,h^{-1} \times 10^{11} M_{\odot}/(h^{-1}\mathrm{Mpc})^3\,. \tag{6.6}$$

Suddenly this doesn't look so small. In fact, 10^{11} to 10^{12} solar masses is about the mass of a typical galaxy, and a megaparsec more or less the typical galaxy separation, so the Universe cannot be far away (within an order of magnitude or so) from the critical density. Its density really must be around $10^{-26}\,\mathrm{kg\,m}^{-3}$.

Be sure to understand that the critical density is not necessarily the true density of the Universe, since the Universe need not be flat. However, it sets a natural scale for the density of the Universe. Consequently, rather than quote the density of the Universe directly, it is often useful to quote its value relative to the critical density. This dimensionless quantity is known as the **density parameter** Ω, defined by

$$\Omega(t) \equiv \frac{\rho}{\rho_c} . \tag{6.7}$$

Again, in general Ω is a function of time, since both ρ and ρ_c depend on time. The present value of the density parameter is denoted Ω_0.

With this new notation, we can rewrite the Friedmann equation in a very useful form. Substituting in for ρ in equation (6.3) using the definitions I have made, equations (6.4) and (6.7), leads to

$$H^2 = \frac{8\pi G}{3} \rho_c \, \Omega - \frac{k}{a^2} = H^2 \, \Omega - \frac{k}{a^2} , \tag{6.8}$$

and rearranging gives

$$\Omega - 1 = \frac{k}{a^2 H^2} . \tag{6.9}$$

We see that the case $\Omega = 1$ is very special, because in that case k must equal zero and since k is a fixed constant this equation becomes $\Omega = 1$ for all time. That is true independent of the type of matter we have in the Universe, and this is often called a **critical-density Universe**. When $\Omega \neq 1$, this form of the Friedmann equation is very useful for analyzing the evolution of the density, as we will see later in the chapter on inflationary cosmology.

Our Universe contains several different types of matter, and this notation can be used not just for the total density but also for each individual component of the density, so one talks of Ω_{mat}, Ω_{rad} etc. Some cosmologists even define a 'density parameter' associated with the curvature term, by writing

$$\Omega_k \equiv -\frac{k}{a^2 H^2} . \tag{6.10}$$

This can be positive or negative, and using it the Friedmann equation can be written as

$$\Omega + \Omega_k = 1 . \tag{6.11}$$

We'll return to the observational status of Ω_0 in Chapter 9.

6.3 The deceleration parameter q_0

As we've discovered, not only is the Universe expanding, but also the rate at which it is expanding, given by the Hubble parameter, is changing with time. The deceleration parameter is a way of quantifying this.

Consider a Taylor expansion of the scale factor about the present time. The general form of this (with dots as always indicating time derivatives) is

$$a(t) = a(t_0) + \dot{a}(t_0)\,[t - t_0] + \frac{1}{2}\ddot{a}(t_0)\,[t - t_0]^2 + \cdots . \tag{6.12}$$

Let's divide through by $a(t_0)$. Then the coefficient of the $[t - t_0]$ term will just be the present Hubble parameter, and we can write

$$\frac{a(t)}{a(t_0)} = 1 + H_0\,[t - t_0] - \frac{q_0}{2} H_0^2\,[t - t_0]^2 + \cdots , \tag{6.13}$$

which defines the deceleration parameter q_0 as

$$q_0 = -\frac{\ddot{a}(t_0)}{a(t_0)}\frac{1}{H_0^2} = -\frac{a(t_0)\ddot{a}(t_0)}{\dot{a}^2(t_0)} . \tag{6.14}$$

The larger the value of q_0, the more rapid the deceleration.

The simplest situation is if the Universe is matter dominated, $p = 0$. Remember that by 'matter' we mean any pressureless material; it could be a collection of elementary particles, or equally well a collection of galaxies. Then from the acceleration equation (3.18) and the definition of critical density, equation (6.4), we find

$$q_0 = \frac{4\pi G}{3}\rho\,\frac{3}{8\pi G \rho_c} = \frac{\Omega_0}{2} . \tag{6.15}$$

So in this case, a measurement of q_0 would immediately tell us Ω_0.

If we know the properties of the matter in the Universe, then q_0 is not independent of the first two parameters we have discussed, H_0 and Ω_0. Those two are sufficient to describe all the possibilities. However, we don't know everything about the material in the Universe, so q_0 can provide a new way of looking at the Universe. It can in principle be measured directly by making observations of objects at very large distances, such as the numbers of distant galaxies, because the deceleration governs how large the Universe would be at an earlier time.

Recently, the first convincing measurements of q_0 have been made by two research groups studying distant supernovae of a class known as type Ia, which are believed to be good standard candles. To widespread surprise, the result is that the Universe appears to be *accelerating* at present, $q_0 < 0$.[1] *None* of the cosmological models that we have discussed so far are capable of satisfying this condition, as can be seen directly from the acceleration equation (3.18). This result is becoming firmly established, and is amongst the most dramatic observational results in modern cosmology. The following chapter discusses how to extend our simple cosmological models to account for it.

---

[1] The mathematical tools required to analyze such data are beyond the scope of the main body of this book, but are described in Advanced Topic 2, where the supernova observations are discussed in greater detail.

Problems

6.1. The deceleration parameter is defined by equation (6.14). Use the acceleration equation (3.18) and the definition of critical density to show that a radiation-dominated Universe has $q_0 = \Omega_0$.

6.2. Identify a sufficient and necessary condition that must be satisfied by the equation of state if q_0 is to be negative.

Chapter 7

The Cosmological Constant

7.1 Introducing Λ

When formulating general relativity, Einstein believed that the Universe was static, but found that his theory of general relativity did not permit it. This is simply because all matter attracts gravitationally; none of the solutions we have found correspond to a static Universe with constant a. In order to arrange a static Universe, he proposed a change to the equations, something he would later famously call his "greatest blunder". That was the introduction of a cosmological constant.

The introduction of such a term is permitted by general relativity, and although Einstein's original motivation has long since faded, it is currently seen as one of the most important and enigmatic objects in cosmology. The cosmological constant Λ appears in the Friedmann equation as an extra term, giving

$$H^2 = \frac{8\pi G}{3}\rho - \frac{k}{a^2} + \frac{\Lambda}{3}. \tag{7.1}$$

Here Λ has units $[\text{time}]^{-2}$, though some people include an explicit factor of c^2 in this equation to instead measure it as $[\text{length}]^{-2}$.

In principle, Λ can be positive or negative, though the positive case is much more commonly considered. Einstein's original idea was to balance curvature, Λ and ρ to get $H(t) = 0$ and hence a static Universe (see Problem 7.2). In fact, this idea was rather misguided, since such a balance proves to be unstable to small perturbations, and hence presumably couldn't arise in practice. Nowadays, the cosmological constant is most often discussed in the context of Universes with the flat Euclidean geometry, $k = 0$.

The effect of Λ can be seen more directly from the acceleration equation. Following the derivation of Section 3.5, but now using the Friedmann equation as given above, gives

$$\frac{\ddot{a}}{a} = -\frac{4\pi G}{3}\left(\rho + \frac{3p}{c^2}\right) + \frac{\Lambda}{3}. \tag{7.2}$$

A positive cosmological constant gives a positive contribution to $\ddot{a}$, and so acts effectively as a repulsive force. In particular, if the cosmological constant is sufficiently large, it can overcome the gravitational attraction represented by the first term and lead to an accelerating Universe. It can therefore explain the observed acceleration of the Universe described in Section 6.3.

In the same way that it is useful to express the density as a fraction of the critical density, it is convenient to define a density parameter for the cosmological constant as

$$\Omega_\Lambda = \frac{\Lambda}{3H^2} \,. \tag{7.3}$$

Although Λ is a constant, Ω_Λ is not since H varies with time. Repeating the steps used to write the Friedmann equation in the form of equation (6.9), we then find

$$\Omega + \Omega_\Lambda - 1 = \frac{k}{a^2 H^2} \,. \tag{7.4}$$

The condition to have a flat Universe, $k = 0$, generalizes to

$$\Omega + \Omega_\Lambda = 1 \,. \tag{7.5}$$

The usual convention amongst astronomers, which I will follow in this book, is that the cosmological constant term is not considered to be part of the matter density Ω. (Particle physicists, on the other hand, often include the cosmological constant as one of the components of the total density.) The relation between the density parameters and the geometry now becomes

Open Universe:	$0 < \Omega + \Omega_\Lambda < 1$.
Flat Universe:	$\Omega + \Omega_\Lambda = 1$.
Closed Universe:	$\Omega + \Omega_\Lambda > 1$.

7.2 Fluid description of Λ

It is often helpful to describe Λ as if it were a fluid with energy density ρ_Λ and pressure p_Λ. From equation (7.1), we see that the definition

$$\rho_\Lambda \equiv \frac{\Lambda}{8\pi G} \tag{7.6}$$

brings the Friedmann equation into the form

$$H^2 = \frac{8\pi G}{3}(\rho + \rho_\Lambda) - \frac{k}{a^2} \,. \tag{7.7}$$

This definition also ensures that $\Omega_\Lambda \equiv \rho_\Lambda/\rho_c$, where ρ_c is the critical density.

In order to determine the effective pressure corresponding to Λ, one can seek a definition so that the acceleration equation with Λ reduces to its standard form, equation (3.18), with $\rho \to \rho + \rho_\Lambda$ and $p \to p + p_\Lambda$. More directly, we can consider the fluid equation for Λ

$$\dot{\rho_\Lambda} + 3\frac{\dot{a}}{a}\left(\rho_\Lambda + \frac{p_\Lambda}{c^2}\right) = 0\,. \tag{7.8}$$

Since ρ_Λ is constant by definition, we must have

$$p_\Lambda = -\rho_\Lambda c^2\,. \tag{7.9}$$

The cosmological constant has a *negative* effective pressure. This means that as the Universe expands, work is done on the cosmological constant fluid. This permits its energy density to remain constant even though the volume of the Universe is increasing.

Concerning its physical interpretation, Λ is sometimes thought of as the energy density of 'empty' space. In particular, in quantum physics one possible origin is as a type of 'zero-point energy', which remains even if no particles are present, though unfortunately particle physics theories tend to predict that the cosmological constant is far larger than observations allow. This discrepancy is known as the **cosmological constant problem**, and is one of the key unsolved problems in elementary particle physics.

It may be that the cosmological constant is only a transient phenomenon, which will disappear in the future. Another possibility, often called **quintessence**, is that the cosmological constant is not actually perfectly constant but exhibits slow variation. For instance, one could assume the quintessence 'fluid' to have equation of state

$$p_Q = w\rho_Q c^2\,, \tag{7.10}$$

where w is a constant. The case $w = -1$ corresponds to a cosmological constant, while more generally accelerated expansion is possible provided $w < -1/3$ (you explored some solutions of this type in Problem 5.3). However in this book I will only consider the case of a perfect cosmological constant.

7.3 Cosmological models with Λ

The introduction of Λ has forced cosmologists to rethink some of the standard lore of cosmology, as it greatly increases the range of possible behaviours of the Universe. For instance, it is no longer necessarily true that a closed Universe ($k > 0$) recollapses, nor that an open Universe expands forever. In fact, if the cosmological constant is powerful enough, there need not even be a Big Bang, with the Universe instead beginning in a collapsing phase, followed by a bounce at finite size under the influence of the cosmological constant (though such models are ruled out by observations). It is also possible to have a prolonged phase where the Universe remains almost static, known as 'loitering', by arranging parameters so that the Universe closely approaches the unstable Einstein static Universe.

As the Hubble parameter only provides an overall scaling factor, a useful way to parametrize possible models is to focus on the two other parameters, the present densities of matter and of the cosmological constant. An excellent assumption is to assume the matter in the present Universe is pressureless. Different models can then be identified by

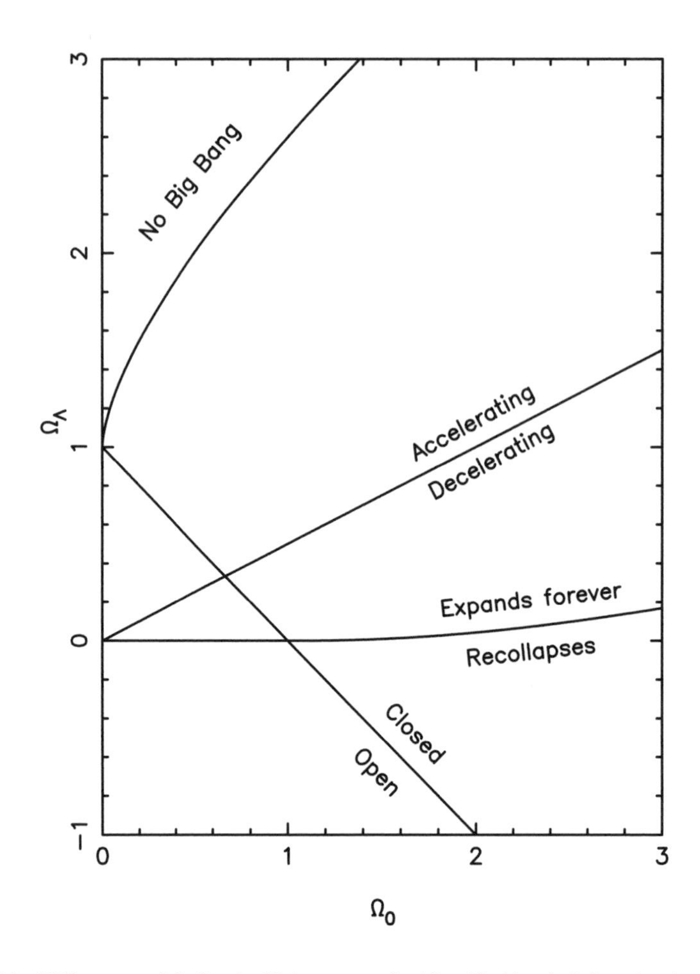

Figure 7.1 Different models for the Universe can be identified by their location in the plane showing the densities of matter and Λ. This figure indicates the main properties in different regions, with the labels indicating the behaviour on each side of the dividing lines.

their location in the plane of Ω_0 and Ω_Λ, as shown in Figure 7.1.[1] We have already seen that the line $\Omega_0 + \Omega_\Lambda = 1$ gives a flat Universe, and divides the plane into open and closed cosmologies.

To identify where in the plane we have an accelerating Universe, we need an expression for the deceleration parameter q_0. A pressureless Universe with a cosmological constant has

$$q_0 = \frac{\Omega_0}{2} - \Omega_\Lambda \,, \tag{7.11}$$

which you are asked to derive in Problem 7.3, and so we have acceleration provided $\Omega_\Lambda > \Omega_0/2$. If we additionally assume that the geometry is flat, this relation simplifies

[1]Beware the somewhat sloppy notation of sometimes using Ω_Λ to indicate the present value of this quantity.

further to $q_0 = 3\Omega_0/2 - 1$, and we have acceleration if $\Omega_\Lambda > 1/3$.

The other two main properties are whether there is a Big Bang, and whether the Universe will eventually collapse or not. There are analytic expressions for these curves, shown in Figure 7.1, but they are too complicated to give here. For $\Omega_0 \leq 1$, whether there is recollapse or not depends simply on the sign of Λ, but for $\Omega_0 > 1$ the gravitational attraction of matter can overcome a small positive cosmological constant and cause recollapse.

While most cosmologists would have preferred the cosmological constant to equal zero, the Universe itself appears to have other ideas, with the observations of distant type Ia supernovae mentioned at the end of the previous chapter arguing strongly in favour of a presently-accelerating Universe. The observations leading to that conclusion are explored in more detail in Advanced Topic 2.3, and even if you are not planning to read that section I suggest you have a look at Figure A2.4 on page 132 which shows the observational constraints superimposed on the Ω_0–Ω_Λ plane. These observations demand inclusion of the cosmological constant; it is now regarded as an essential part of cosmological models aiming to explain observational data, and understanding its value is one of the mysteries of fundamental physics.

Problems

7.1. Suppose that the Universe contains four different contributions to the Friedmann equation, namely radiation, non-relativistic matter, a cosmological constant, and a negative (hyperbolic) curvature. Write down the way in which each of these terms behaves as a function of the scale factor $a(t)$. Which of them would you expect to dominate the Friedmann equation at early times, and which at late times?

7.2. By considering both the Friedmann and acceleration equations, and assuming a pressureless Universe, demonstrate that in order to have a static Universe we must have a closed Universe with a positive vacuum energy. Using either physical arguments or mathematics, demonstrate that this solution must be unstable.

7.3. Confirm the result, quoted in the main text, that a pressureless Universe with a cosmological constant has a deceleration parameter given by

$$q_0 = \frac{\Omega_0}{2} - \Omega_\Lambda(t_0)\,.$$

7.4. The most likely cosmology describing our own Universe has a flat geometry with a matter density of $\Omega_0 \simeq 0.3$ and a cosmological constant with $\Omega_\Lambda(t_0) \simeq 0.7$. What will the values of Ω and Ω_Λ be when the Universe has expanded to be five times its present size? Use an approximation suggested by this result to find the late-time solution to the Friedmann equation for our Universe. What is the late-time value of the deceleration parameter q?

7.5. Show that in a spatially-flat matter-dominated cosmology the density parameter evolves as

$$\Omega(z) = \Omega_0 \frac{(1+z)^3}{1 - \Omega_0 + (1+z)^3 \Omega_0} .$$

If our Universe has $\Omega_0 \simeq 0.3$, at what redshift did it begin accelerating?

Chapter 8

The Age of the Universe

One of the quantities that we can predict from a cosmological model, from the solution $a(t)$ for the expansion, is the age of the Universe t_0. This offers an opportunity to connect the age of the Universe itself with the ages of objects within it, though gaps in our cosmological knowledge leave the situation somewhat uncertain. Historically there has been much concern as to whether the predicted age of the Universe is large enough to accommodate the ages of its contents, but in recent years such fears have largely disappeared.

Let's start with an approximate estimation. The characteristic rate of expansion is given by the Hubble parameter, so a first guess at the age of the Universe is that it is the timescale associated with the Hubble parameter, namely H_0^{-1}, since $r = vH_0^{-1}$ is basically distance equals velocity multiplied by time. This estimate is approximate, because it ignores the fact that v changes with time under the effect of gravity. We've been writing the Hubble constant as

$$H_0 = 100h \,\mathrm{km\,s^{-1}\,Mpc^{-1}}\,, \tag{8.1}$$

which isn't quite what we need, because the unusual units conceal the fact that H simply has the units of $[\mathrm{time}]^{-1}$. We have to convert the kilometers into megaparsecs (or vice versa) to cancel them out, and then convert seconds to years. What we get, using values quoted on page xiv, is

$$H_0^{-1} = 9.77\, h^{-1} \times 10^9 \text{ yrs}\,. \tag{8.2}$$

This is known as the **Hubble time**. The uncertainty in the Hubble constant, indicated by h, means we have to tolerate an uncertainty in this crude estimate of the age of the Universe. But the message is that ten billion years is a good first guess at the age of the Universe.

Before progressing to a better calculation, let's find out what observations tell us of the age of the Universe. The geological timescale gives us a good estimate of the age of the Earth, which is about five billion years. But it is not thought that the Earth is nearly as old as the Universe. There are various ways of dating other objects in the Universe. The relative amounts in the galactic disk of Uranium isotopes, which have lifetimes comparable to the age of the Universe, suggests an age around ten billion years — see Problem 8.1. Estimates consistent with this are also found from studying the cooling of white dwarf stars after they form. However, the best method is thought to be to use the chemical evolution of stars in old globular clusters, which are believed to be amongst the oldest objects in the Universe. Until 1997 these were thought to be alarmingly old, but then the Hipparcos satellite discovered that nearby stars are further away than we thought, hence brighter, hence burning fuel faster and hence younger. The best estimate now is an age in the range ten to thirteen billion years, with perhaps an extra billion years to be added in before they manage to form in the first place.

It should be stressed that it is remarkable that the cosmological theory and the age measurements are in the same ballpark at all. That in itself is a strong vindication that the ideas behind the Big Bang are along the right track. Our crude estimate above seems to adequately account for the observed ages of objects in the Universe.

What happens if we try to do better with our theoretical estimates? Although the precise cosmological model describing our Universe is uncertain, we are pretty sure that it has been matter dominated (i.e. dominated by some form of pressureless material) for some considerable time, and so we can use the matter-dominated evolution to calculate the age. Let's first suppose that the Universe has the critical density. Then we already have the solution, equation (5.15), which is

$$a(t) = \left(\frac{t}{t_0}\right)^{2/3} \quad \Longrightarrow \quad H \equiv \frac{\dot{a}}{a} = \frac{2}{3t}, \tag{8.3}$$

and so the present Hubble parameter is

$$H_0 = \frac{2}{3t_0}. \tag{8.4}$$

So in such a Universe, the age is actually shorter than our naïve estimate — it's only

$$t_0 = \frac{2}{3}H_0^{-1} = 6.51\,h^{-1} \times 10^9\ \text{yrs}. \tag{8.5}$$

The extra factor of 2/3 has removed much of the room for comfort; if h is towards the top of its measured range then we don't get up to the ten billion years or so required by observations. While $h = 0.6$ gives a marginally-acceptable eleven billion years, if $h = 0.8$ we predict an age of only eight billion years.

What can happen to reconcile this? Well, if the Universe is closed then the age becomes even less and the situation is becoming very problematic indeed. This is one of many arguments going against the idea of a closed Universe.

On the other hand, moving to an open Universe with $\Omega_0 < 1$ helps. The physical interpretation is that if there is less matter, then it would have taken longer for the gravitational attraction to slow the expansion to its present rate. That last sentence needs a bit of thought before it sinks in. Try thinking of two trains travelling at 100 miles per hour (or your favourite metric unit); they both start to brake and you ask how long before they slow down to 50 miles per hour. The one with the inferior brakes takes longer to do so. The same works for the Universe; if there is less matter it requires longer for the gravitational deceleration to slow it down to the observed expansion rate. The detailed result is studied in Problem 8.2.

In the limit $\Omega_0 \to 0$ there is no gravity at all, and hence no deceleration, so the esti-

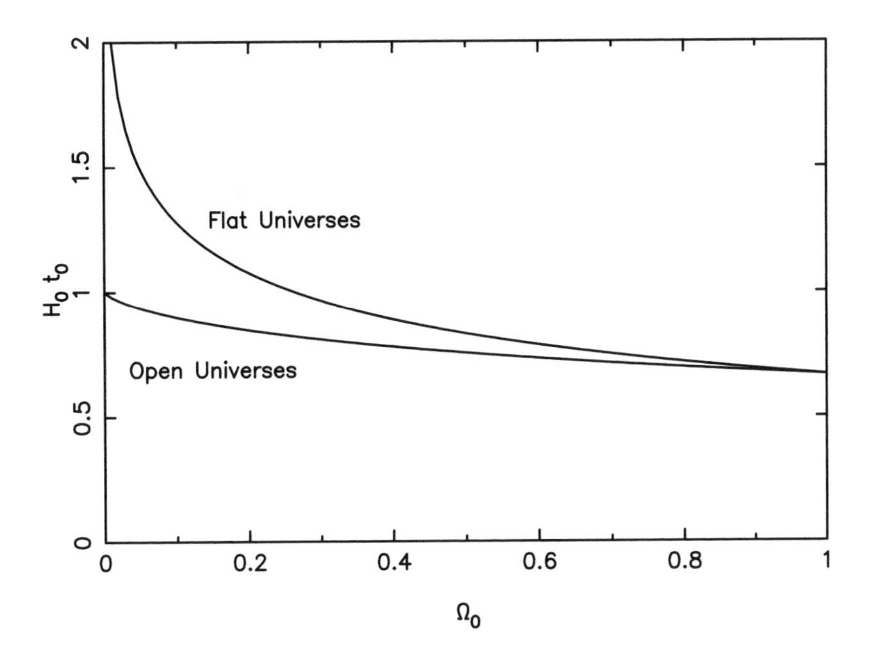

Figure 8.1 Predicted ages as fractions of the Hubble time H_0^{-1}, for open Universes and for Universes with a flat geometry plus a cosmological constant. The prediction $H_0 t_0 = 2/3$ for critical density models is at the right-hand edge.

mate of equation (8.2), based on assuming constant velocity, becomes correct and we get $t_0 = H_0^{-1}$. However, clearly we can't countenance a Universe with no matter at all in it, so we can only get part of the way there.

Observations suggest that a better option for a low-density Universe is to retain the flat geometry by introducing a positive cosmological constant. As this opposes the deceleration, it has a more severe effect in increasing the age than going to an open Universe, and if the density is low enough we can get an age which exceeds H_0^{-1}. Deriving the formula for the age is tricky (see Problem 8.4), but for reference there are two equivalent (and equally unpleasant) forms

$$H_0 t_0 = \frac{2}{3}\frac{1}{\sqrt{1-\Omega_0}} \ln\left[\frac{1+\sqrt{1-\Omega_0}}{\sqrt{\Omega_0}}\right] = \frac{2}{3}\frac{1}{\sqrt{1-\Omega_0}} \sinh^{-1}\left[\sqrt{\frac{1-\Omega_0}{\Omega_0}}\right] . \quad (8.6)$$

The 'break-even' point where $t_0 = H_0^{-1}$ is at $\Omega_0 = 0.26$, close to the value preferred by observation. For a given value of h, the cosmological constant gives us the oldest age, and for the favoured values $\Omega_0 \simeq 0.3$ and $h \simeq 0.72$ we get an age of about fourteen billion years. This sits comfortably with the estimated ages given earlier.

Figure 8.1 shows the predicted ages for these different cosmological models, as fractions of the Hubble time H_0^{-1}. For $\Omega_0 = 1$ we get an age which is two-thirds of the Hubble time, according to equation (8.5), while for lower densities the age becomes older, assuming H_0 is kept fixed.

Problems

8.1. The galaxy's age can be estimated by radioactive decay of Uranium. Uranium is produced as an r-process element in supernovae (don't worry if you don't know what that is!), and on this basis the initial abundances of the two isotopes U^{235} and U^{238} are expected to be in the ratio

$$\left.\frac{U^{235}}{U^{238}}\right|_{\rm initial} \simeq 1.65\,.$$

The decay rates of the isotopes are

$$\lambda(U^{235}) = 0.97 \times 10^{-9}\,{\rm yr}^{-1}\,;$$

$$\lambda(U^{238}) = 0.15 \times 10^{-9}\,{\rm yr}^{-1}\,.$$

Finally, the present abundance ratio is

$$\left.\frac{U^{235}}{U^{238}}\right|_{\rm final} \simeq 0.0072\,.$$

Use the decay law

$$U(t) = U(0)\exp\left(-\lambda t\right)\,,$$

to estimate the age of the galaxy.

Assuming the galaxy took a minimum of an additional billion years to form in the first place, obtain an upper limit on the value of the Hubble parameter h assuming a critical-density Universe.

8.2. In a matter-dominated open Universe, the present age of the Universe is given by the intimidating formula

$$H_0 t_0 = \frac{1}{1-\Omega_0} - \frac{\Omega_0}{2(1-\Omega_0)^{3/2}}\cosh^{-1}\left(\frac{2-\Omega_0}{\Omega_0}\right)\,.$$

[This is the formula which gives the innocent-looking lower curve in Figure 8.1.] Demonstrate that in the limiting case of an empty Universe $\Omega_0 \rightarrow 0$ we get $H_0 t_0 = 1$, and in the limiting case of a flat Universe $\Omega_0 \rightarrow 1$ we recover the result $H_0 t_0 = 2/3$.

[Useful formulae: $\cosh^{-1}(x) \simeq \ln(2x)$ for large x,
$\cosh^{-1}[(1+x)/(1-x)] \simeq 2\sqrt{x} + 2x^{3/2}/3$ for small x.]

8.3. Give a physical argument explaining why introducing a positive cosmological constant will increase the predicted age of the Universe.

8.4. *[For the mathematically-keen only!]* Derive one version of equation (8.6), which gives the age of a spatially-flat cosmology with a cosmological constant. As a first step towards this, demonstrate that the Friedmann equation can be written as

$$\dot{a}^2 = H_0^2 \left[\Omega_0 a^{-1} + (1 - \Omega_0) a^2 \right] ,$$

where a is normalized to be one at the present. Then change the integration variable to a in the expression $t_0 = \int_0^{t_0} dt$.

Chapter 9

The Density of the Universe and Dark Matter

The total density of matter in the Universe is quantified by the density parameter Ω_0. We would like to know not only its value, but also how that density is divided up amongst the different types of material present in our Universe.

9.1 Weighing the Universe

The characteristic scale for the density in the Universe is the critical density ρ_c. As we saw on page 47, it is not a particularly imposing number; its present value is

$$\rho_c = 1.88\,h^2 \times 10^{-26}\,\mathrm{kg\,m^{-3}} = 2.78\,h^{-1} \times 10^{11}\frac{M_\odot}{(h^{-1}\,\mathrm{Mpc})^3}\,. \tag{9.1}$$

An obstacle to comparing the true density to the critical density is the factors of h, which are uncertain. Nevertheless, to get an idea of what is going on, all we have to do is estimate how much material there is in the Universe. From the crude estimates that a typical galaxy weighs about $10^{11}M_\odot$ and that galaxies are typically about a megaparsec apart, we know that the Universe cannot be a long way from the critical density. But how good an estimate can be made?

9.1.1 Counting stars

The simplest thing we can do is look at all the stars within a suitably-large region. Stellar structure theory gives a good estimate of how massive a star is for a given temperature and luminosity. Provided we have looked in a large enough region, we get an estimate of the overall density of material in stars. This has been done by many researchers, and the answer obtained is that the density in stars is a small fraction of the critical density, around

$$\Omega_{\mathrm{stars}} \equiv \frac{\rho_{\mathrm{stars}}}{\rho_c} \simeq 0.005 \rightarrow 0.01\,. \tag{9.2}$$

Notice that this number is independent of h, even though the critical density depends on h^2. That is because the estimate is carried out by adding up the light flux; since distances are uncertain by a factor h and the light flux falls off as the square of the distance, the h dependence cancels out of the final answer.

9.1.2 Nucleosynthesis foreshadowed

Not all of the material we are able to see is in the form of stars. For example, within clusters of galaxies there is a large amount of gas which is extremely hot and emits in the X-ray region of the spectrum, which I will discuss further below. Another possibility is that a lot of material resides in very low mass stars, which would be too faint to detect. Often discussed are brown dwarfs (sometimes called Jupiters), which are 'stars' with insufficient material to initiate nuclear burning. Objects with mass less than $0.08M_\odot$ are thought to be in this class. If for some reason there were a lot of objects of this kind then they could contribute substantially to the total density without being noticed, though this is not thought to be very likely on grounds of extrapolation from what we do know.

Nevertheless, there is a very strong reason to believe that conventional material cannot contribute an entire critical density. That evidence comes from the theory of nucleosynthesis — the formation of light elements — which will be discussed in Chapter 12. This theory can only match the observed element abundances if the amount of *baryonic* matter has a density

$$0.016 \leq \Omega_{\mathrm{B}} h^2 \leq 0.024\,. \tag{9.3}$$

Recall from Section 2.5 that baryonic matter means protons and neutrons, and hence refers to the kinds of particle that we and our environment are made from.

In this expression the Hubble constant appears as an additional uncertainty, but the constraint is certainly strong enough to insist that it is not possible to have an entire critical density worth of baryonic matter, whether it be in the form of luminous stars or invisible brown dwarfs or gas. Adopting the Hubble Space Telescope constraints on h gives an upper limit well below ten percent.

Nucleosynthesis also gives a lower bound on Ω_{B} which suggests that there should be substantially more baryonic material in the Universe than just the visible stars, probably upwards of 2.5 percent of the critical density. This is in good agreement with observations of galaxy clusters discussed below.

9.1.3 Galaxy rotation curves

In fact, there is considerable dynamical evidence that there is more than just the visible matter. The history of this subject is surprisingly old; in 1932 Oort found evidence for extra hidden matter in our galaxy, and one year later Zwicky inferred a large density of matter within clusters of galaxies, a result which has stood the test of time extremely well. The general argument is to look at motions of various kinds of astronomical object, and assess whether the visible material is sufficient to provide the inferred gravitational force. If it is not, the excess gravitational attraction must be due to extra, invisible, material.

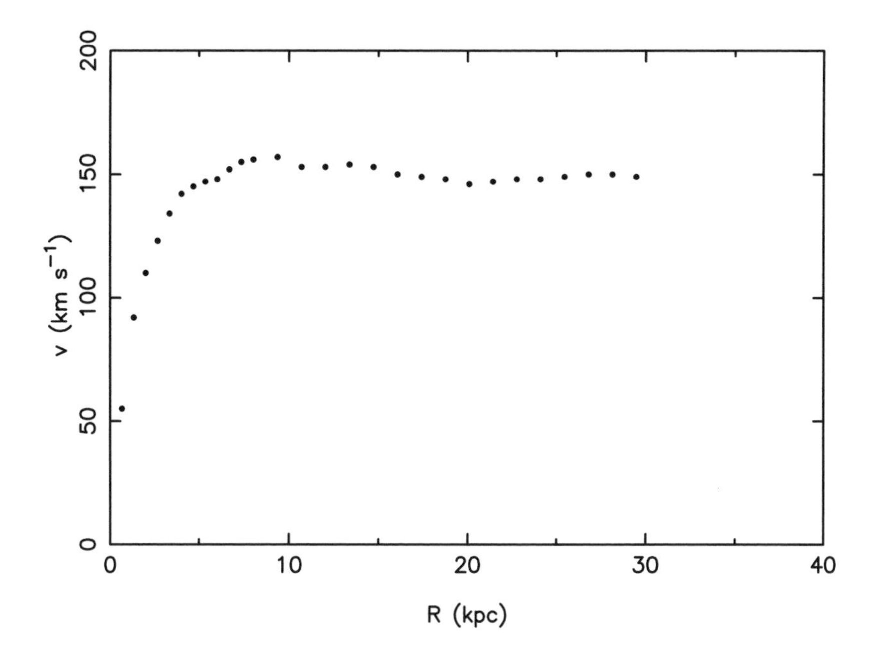

Figure 9.1 The rotation curve of the spiral galaxy NGC3198. We see that it remains roughly constant at large radii, outside the visible disk. Faster than expected orbits require a larger central force, and so they imply the existence of extra, dark, matter.

One of the most impressive applications of this simple idea is to galaxy rotation curves. A galaxy rotation curve shows the velocity of matter rotating in a spiral disk, as a function of radius from the center. The individual stars are on orbits given by Kepler's law; if a galaxy has mass $M(R)$ within a radius R, then the balance between the centrifugal acceleration and the gravitational pull demands that its velocity obeys

$$\frac{v^2}{R} = \frac{GM(R)}{R^2}, \tag{9.4}$$

which can be rewritten as

$$v = \sqrt{\frac{GM(R)}{R}}. \tag{9.5}$$

The mass outside the radius R contributes no gravitational pull, due to the same theorem of Newton's we used to derive the Friedmann equation in Chapter 3.

At large distances, enclosing most of the visible part of the galaxy, we expect the mass to be roughly constant and so the rotational velocity should drop off as the square root of R. At such large distances, the rotation is mapped out by interstellar gas, and instead is found to stay more or less constant, as shown in Figure 9.1.[1] The typical velocities at large radii can be three times higher than predicted from the luminous matter, implying ten

[1]Note that it is the velocity itself, and not the angular velocity, which is constant, so the galaxy is still rotating differentially and certainly not as a rigid body.

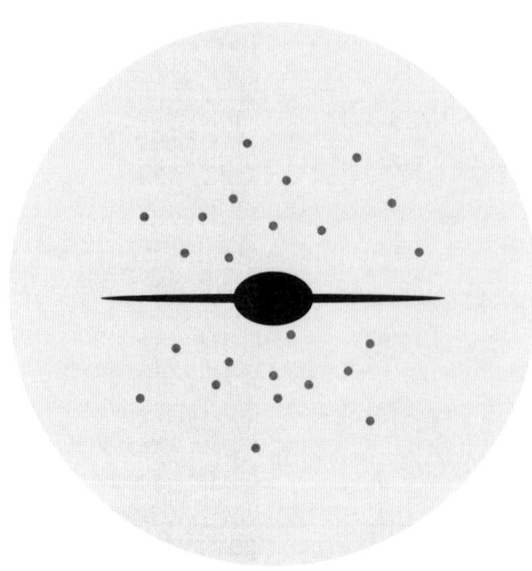

Figure 9.2 A schematic illustration of a galactic disk, with a few globular clusters, embedded in a spherical halo of dark matter.

times more matter than can be directly seen. This is an example of **dark matter**. Standard estimates suggest

$$\Omega_{\text{halo}} \simeq 0.1\,. \tag{9.6}$$

It is just about possible given present observations that this matter can be entirely baryonic, since this is marginally consistent with equation (9.3). However, many models based on low mass stars and/or brown dwarfs have been excluded and it is probably difficult to make up all of the halo with them. A popular alternative is to suggest that this density is in some new form of matter, which is non-baryonic and only interacts extremely weakly with conventional matter. This is reinforced by higher estimates for the matter density on larger scales discussed next. It is usually assumed that this dark matter lacks any dissipation mechanism able to concentrate it into a disk structure resembling that of the stars. If that is the case, then the dark matter should be in the form of a spherical halo, meaning a solid sphere with high density at the centre falling off to smaller values at large radii. The visible galactic disk and the globular clusters are embedded in this halo, as shown in Figure 9.2.

9.1.4 Galaxy cluster composition

Galaxy clusters are the largest gravitationally-collapsed objects in the Universe, and as such are an ideal probe of the different kinds of matter. Because of their size, they should contain a fair sample of the material in the Universe, since there is no means of segregating different types of material as all is drawn in by gravity. The visible components of a galaxy cluster are in two main parts, seen in Figure 2.3 on page 6. There are the stars within the individual galaxies, and there is diffuse hot gas seen in X-rays which has been heated up through falling into the strong gravitational potential well of the cluster. It turns out that the baryon content of the latter is the greater, with about five to ten times more hot gas than stars. While only a small fraction of galaxies are in clusters, galaxy cluster formation is an ongoing process and those still to form ought to resemble those already present. The simplest assumption is that there is a considerable amount of cool gas between galaxies, which present technology does not allow us to detect, which will be heated if it becomes incorporated into a galaxy cluster. Note that the relative amounts of stars and hot gas in galaxy clusters are in excellent agreement with a comparison of the observed density of stars to the total baryon density inferred from nucleosynthesis.

The hot gas can also be used to estimate the amount of dark matter present. Its high temperature gives it a substantial pressure, but it is confined to the galaxy cluster by gravitational attraction. However, the self-gravity of the gas alone does not provide enough attraction on its own, with the total mass of the cluster inferred to be around ten times larger than the gas mass. It is natural to assume that this extra attraction is given by dark matter, and if so the dark matter density must be around ten times larger than the baryon density given by nucleosynthesis. For example, data from the *Chandra* X-ray satellite have been used to give

$$\frac{\Omega_{\rm B}}{\Omega_0} \simeq 0.065\, h^{-3/2}\,, \tag{9.7}$$

which using the nucleosynthesis constraint on $\Omega_{\rm B}$ given above leads to

$$\Omega_0 \simeq 0.3\, h^{-1/2} \simeq 0.35\,. \tag{9.8}$$

This type of analysis indicates very directly that the matter density is dominated by dark matter, but that the dark matter density falls short of the critical density.

9.1.5 Bulk motions in the Universe

Further dynamical evidence for the existence of dark matter comes from the motions of galaxies relative to one another (i.e. the deviations from the cosmological principle). As with the rotation curves, the gravitational force exerted on galaxies by their neighbours depends on the total mass of the galaxies, whether it is visible or not. Galaxies possess relative motions, the peculiar velocities mentioned earlier, which allows one to estimate their mass under the assumption that their gravitational interaction is responsible for the motions, which are often termed bulk flows.

The analyses tend to be rather complicated, and indicate that the total density of matter in the Universe must obey

$$\Omega_0 \geq 0.2\,. \tag{9.9}$$

As with the galaxy cluster determination of Ω_0, this is well above the amount permitted by the nucleosynthesis constraint of equation (9.3). The conclusion therefore is not only that the Universe is largely composed of dark matter, but also that this dark matter must be non-baryonic rather than an invisible form of conventional material.

9.1.6 The formation of structure

One of the most exciting areas of modern cosmology is the study of the origin of structure in the Universe, such as galaxies, galaxy clusters, and irregularities in the cosmic microwave background. As described in Advanced Topic 5, the modern view of the origin of structure is that it grew from initially small irregularities through gravitational attraction, which draws material towards regions which start off with higher than average density. As gravity is the driving force, the formation of structure is a probe of the total density of matter, just like rotation curves and bulk flows.

For upwards of twenty years, it has been widely accepted that the baryonic matter in the Universe would not in itself provide enough gravitational attraction to form the observed structures by the present age of the Universe. This problem can be circumvented by the introduction of non-baryonic dark matter, which provides the extra gravitational force to allow structures to form more quickly and is not inhibited by pressure effects. It is thought that this can only work if Ω_0 obeys

$$\Omega_0 \geq 0.2\,. \tag{9.10}$$

There are presently no working models of structure formation which do not rely on at least this amount of dark matter.

9.1.7 The geometry of the Universe and the brightness of supernovae

One of the most remarkable observations in cosmology in recent years has been the first precision measurement of the geometry of the Universe using structures in the cosmic microwave background. I can only give a flavour of the result here, though more details will be given later in the book. As explained in Advanced Topic 5.4, structure formation scenarios predict a characteristic angular size, of around one degree, for features seen in the microwave background. The precise scale depends primarily on the geometry of the Universe, which tells us how the microwave photons travelled from their origin to our location.

The first precision measurement of the size of these features was announced by the Boomerang experiment in April 2000, followed swiftly by confirmation from the Maxima experiment. While the precise result does depend somewhat on assumptions, the simplest interpretation of those results is that the Universe is close to spatially flat, with the total density (including any cosmological constant) lying within ten percent of the critical density. Further and more accurate confirmation of this result has come from the WMAP satellite, reducing the uncertainty to around one percent [see equation (A5.7)].

This microwave background data are especially powerful when combined with the data on supernova brightness described in Chapter 7. Both can be represented in the Ω_0–Ω_Λ plane. A full discussion is made in the Advanced Topics, but I suggest you look now to Figure A2.4 on page 132. The supernova data cross the line of flat geometry almost at a right angle, and hence the region capable of fitting both data sets is extremely small. The favoured values are $\Omega_0 \simeq 0.3$ and $\Omega_\Lambda \simeq 0.7$, and the former is in excellent agreement with other measures of the matter density given earlier in this section. Note that the combination of direct measures of the dark matter density with the WMAP results gives support to the cosmological constant independently of the supernova observations.

We will later see that the theory of **cosmological inflation**, discussed in Chapter 13, makes the prediction that the Universe has a flat geometry. Cosmologists have long used this theoretical argument to justify them adopting the spatially-flat case, but it is only very recently that it has had direct support from observation.

9.1.8 Overview

To summarize, observational evidence paints a consistent picture as follows:

- Luminous baryonic matter provides less than one percent of the total density.
- Dark baryonic material, probably mostly in the form of cool gas, is the dominant form of baryonic matter, overall making around four percent of the total density.
- There is around ten times as much non-baryonic dark matter as baryonic matter.
- The cosmological constant makes the largest contribution to the total density.
- All components added together give a density equal to, or at least close to, the critical density.

9.2 What might the dark matter be?

The realization that the majority of the matter in the Universe might be non-baryonic is the ultimate Copernican viewpoint; not only are we in no special place in the Universe, but we aren't even made out of the same stuff as dominates the matter density of the Universe. The prediction of non-baryonic dark matter is one of the boldest and most striking in all of cosmology, and if ultimately verified, for example by direct detection of dark matter particles, will be amongst cosmology's most notable successes.

Although the evidence for dark matter is regarded by most as pretty much overwhelming, there is no consensus as to what form it takes. An array of possibilities are discussed below. There are two main classes; in one the dark matter is in the form of individual elementary particles, while in the other it is in some type of compact astrophysical object formed from many particles.

Fundamental particles

- *Things we know exist:* The particle which we know exists and yet whose properties are uncertain enough to allow it to be the dark matter is the neutrino. In the Standard Model of particle interactions the neutrino is a massless particle, and is present in the Universe in great abundance, being about as numerous as photons of light. If the Standard Model is extended to permit the neutrinos to have a small mass (a few tens of electron-volts), this would not affect their number density but they would have enough density to imply a closed Universe! The required density is comparable to, or perhaps slightly higher than, current experimental limits on the electron neutrino, but there are also the neutrinos associated with the muon and tau particles, and so they are more probable candidates. See Problem 9.1 and Advanced Topic 3 for further discussion of this.

A light neutrino would be a type of dark matter known as hot dark matter, meaning that the particles have relativistic velocities for at least some fraction of the Universe's lifetime. In fact, hot dark matter does not have favourable properties for structure formation and if the neutrino has such a mass it is believed that it could at most contribute only part of the matter density, with some other form of dark matter also being required.

Another possibility is that the neutrino could be very heavy, for example comparable to the proton mass. This is allowed because such massive particles wouldn't have as high a number density as photons, since in thermal equilibrium high mass particles are hard to create — the Boltzmann suppression. A heavy neutrino is an example of cold dark matter, meaning particles which have negligible velocities throughout the Universe's history. Having at least some cold dark matter is desirable for structure formation, but a heavy neutrino is much less desirable on particle physics grounds than a light one, and indeed is excluded by particle physics experiments unless the neutrino has unusual properties.

- *Things we believe might exist:* Particle physics theories (particularly those aiming at unification of fundamental forces) have a habit of throwing up all manner of new and as-yet-undiscovered particles, several of which are plausible dark matter candidates. Particle physicists regard **supersymmetry** as the most solidly-founded extension to standard particle theory, and it has the nice property of associating a new companion particle to each of the particles we already know about. In the simplest scenarios, the **lightest supersymmetric particle (LSP)** is stable and is an excellent cold dark matter candidate. Depending on the model the particle in question might be called the photino, or gravitino, or neutralino. They are also sometimes known as WIMPs — Weakly Interacting Massive Particles.

  However the LSP is not the only option for particle dark matter, and other proposed particles include an ultra-light particle known as the axion, ultra-heavy dark matter particles which might for instance be formed at the end of inflation, or particles from a 'shadow Universe' which interacts with our own Universe only gravitationally.

- *Or:* The dark matter could be made of something completely different, which no-one has yet thought of. No-one has ever seen any, after all.

Compact objects

- *Black Holes:* A population of primordial black holes, meaning black holes formed early in the Universe's history rather than at a star's final death throes, would act like cold dark matter. However if they are made of baryons they must form before nucleosynthesis to avoid the nucleosynthesis bound of equation (9.3). Baryons already in black holes by the time of nucleosynthesis don't count as baryons, as they are not available to participate in nuclei formation.

- *MACHOs:* MACHOs are unique amongst dark matter candidates in that they have actually been detected! This rather dubious acronym stands for MAssive Compact

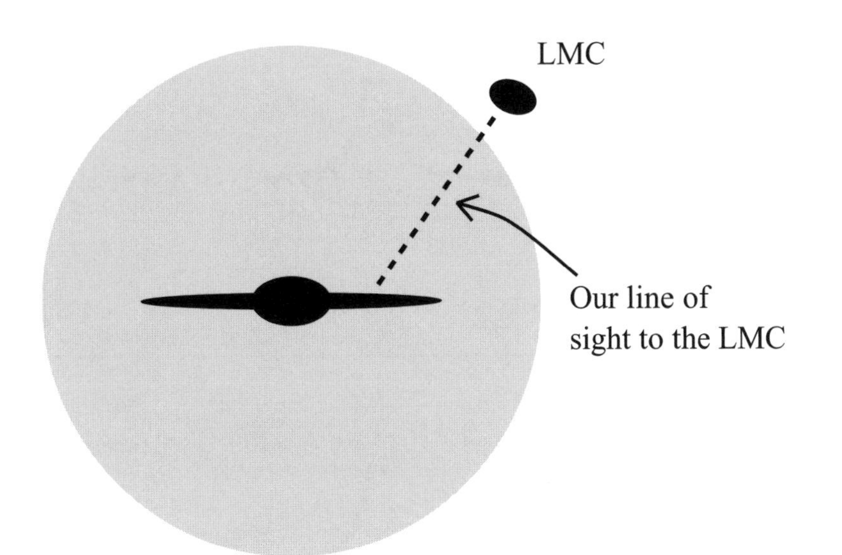

Figure 9.3 An illustration of the MACHO search strategy. We look from our position in the disk of the galaxy towards stars in the Large Magellanic Cloud. The line-of-sight passes through the dark matter halo, and if there are invisible compact objects there, and they pass extremely close to the line-of-sight, then gravitational lensing of the LMC star can occur.

Halo Object, an attempt to counter the WIMP acronym sometimes used for cold dark matter particles. It's a generic term for compact objects with masses not too far from stellar masses, and they may be baryonic or non-baryonic. Brown dwarfs are a baryonic example, but it would also be possible to conceive of non-baryonic examples.

MACHOs have been detected by gravitational lensing of stars in the Large Magellanic Cloud (LMC). The idea, illustrated in Figure 9.3, is to monitor LMC stars, which lie outside (or at least towards the edge of) the galactic halo. If there are invisible massive objects in the halo, and they happen to pass extremely close to our line of sight to the LMC star, then their gravitational field can bend and focus light from the star, temporarily brightening it. The only problem is that such events, called microlensing, are so rare that one has to monitor millions of stars in the LMC, every few days, for a period of years. Impressively, this became possible in the mid 1990s, and, to many people's surprise, MACHOs were detected.

Figure 9.4 shows the brightness of a star in the LMC, monitored for around a year. Such a plot is known as a light curve, and we see a temporary brightening which lasted for around a month. The favoured explanation is gravitational lensing, rather than variability of the star, for several reasons. Firstly, the brightening only happened once, rather than periodically. Secondly, the brightening is the same in both red and blue light, whereas variable stars brighten differently at different wavelengths. And finally, the symmetric shape of the light curves matches that expected if an invisible gravitational lens were to pass in front of the star. The mass of these

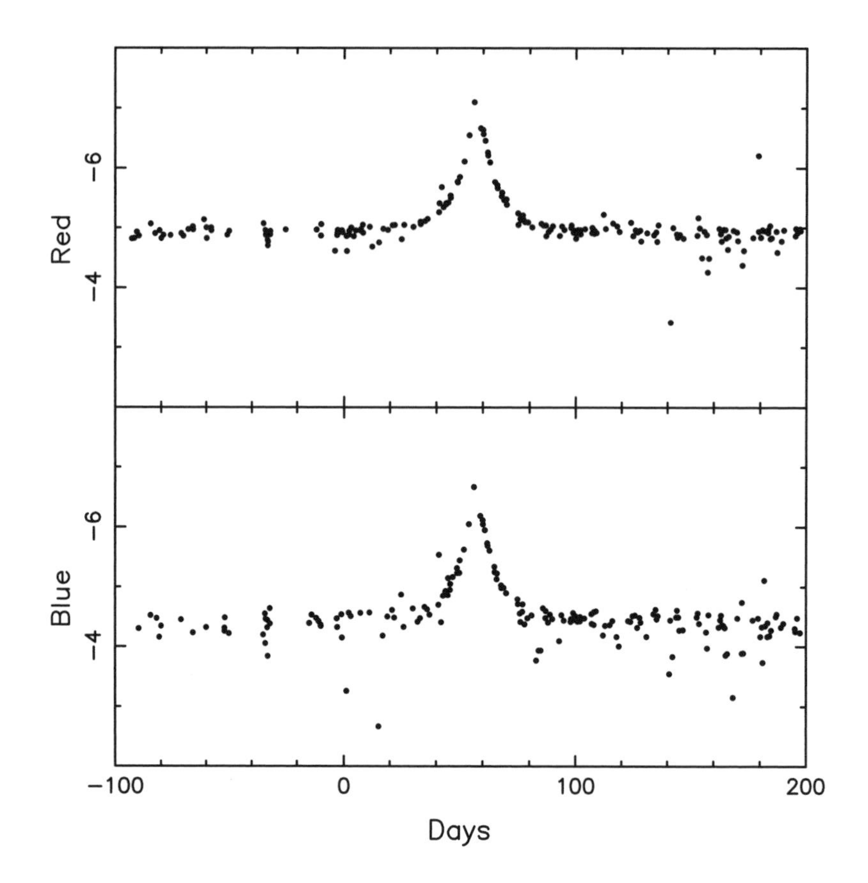

Figure 9.4 Light curves for a star in the LMC, obtained by the MACHO collaboration. The x-axis is in days with an arbitrary origin, while the y-axis shows the brightness of the star in red light and in blue light (the vertical units are a magnitude scale).

invisible objects is estimated as a little less than a solar mass. However, although present they appear to have insufficient density to completely explain the galactic halo. In fact, it may still be possible that both lens and star reside in the LMC itself.

9.3 Dark matter searches

Given the strength of the evidence that most of the matter in the Universe is dark matter, what can be done to discover it and study its properties? We've already discussed the detection of compact dark objects using microlensing, which can be used provided the masses are within a few orders of magnitude of a solar mass. But most of the favoured candidates for non-baryonic dark matter are elementary particles, whose masses are tiny fractions of a gram. Lensing certainly cannot be used to detect these.

The worst case scenario is if the dark matter particles interact with normal matter only through gravitational forces. If that is true then direct detection appears completely impossible; the gravitational force of an individual particle with say a proton mass is minuscule.

The cumulative gravitational force of many such particles is measurable — that's all the astrophysical dark matter evidence I've just discussed — but we want something more tangible.

The best hope is if the dark matter particles interact not only gravitationally, but also through the weak nuclear force (hence the name Weakly Interacting Massive Particle or WIMP).[2] Such interactions could, very reasonably, be feeble enough to have so far remained unobserved, but yet be within the realm of possible detectability. Supersymmetric particles in particular are thought to be potentially detectable if they indeed make up the dark matter.

Remember that the Universe is supposed to be full of these dark matter particles. Consequently, many of them will be streaming through your body at this very instant! Problem 11.2 will investigate this further for the case of neutrinos. You don't notice them because their chance of interacting with you is so slight. But if you collect together enough material, and watch it for long enough, and the interaction rate is high enough, then every so often a dark matter particle will interact with a proton or neutron and give away the secret of its presence.

Many experiments of this type are now in operation around the world. As I write, no detections have been confirmed, but the experimenters live in hope and the sensitivity of the experiments has been improving at an impressive rate. The experiments are challenging, as a typical interaction rate is only once per day per kilogram of material (this applies to your own body every bit as much as to a dark matter detector!). In order to prevent confusion with other interactions, such as cosmic rays or radioactive decay products, a typical experiment is located deep underground. For example, a British collaboration run an experiment 1100 metres underground in the Boulby Mine in Yorkshire. The detectors are further shielded from radioactivity within the mine, and the apparatus cooled to extremely low temperatures. A range of particle detection strategies are used by the different experiments, in order to spot the recoil of atomic nuclei from those rare collisions with dark matter particles.

Picking out a signal from amongst the large range of noise sources is a considerable challenge. One useful check is that the dark matter signal should show an annual modulation; there should be a prevailing flow of dark matter in the solar neighbourhood, and at some parts of its orbit the Earth goes generally in the direction of the flow, decreasing the flux, while at others it goes against the flow and a larger signal should be produced.

[2] Such particles cannot have electromagnetic or strong nuclear interactions, or they would be visible via their direct interactions with conventional matter.

Problems

9.1. As we'll see in the next chapter, radiation in the present Universe, corresponding to a thermal background at 2.725 K, contributes an energy density corresponding to a density parameter of $\Omega_{\rm rad} = 2.47 \times 10^{-5} h^{-2}$. The typical energy of a photon of light in a thermal distribution is given by $3k_{\rm B}T$ [where $k_{\rm B} = 8.6 \times 10^{-5}\,{\rm eV\,K^{-1}}$]. Suppose that the number of neutrinos matches the number of photons. What mass–energy (in electron volts) would these neutrinos have to have in order to contribute a critical density? [Assume that the thermal energy of the neutrinos is negligible compared to their mass–energy.]

The present upper limit on the electron neutrino mass-energy from experiments on Earth is about 10 eV. How low would h have to be to enable electron neutrinos to contribute all the dark matter in a Universe with the critical density?

A more accurate calculation (partly explored in Chapter 11) suggests that the neutrino mass-energy required to give the critical density is larger than the crude calculation above suggests, being about $90h^2$ eV. Is the electron neutrino a realistic dark matter candidate?

9.2. Suppose it were suggested that black holes of mass 10^{-10} solar masses might make up the dark matter in our galactic halo. Make a rough estimate of how far away you'd expect the nearest such black hole to be. How does this compare to the size of the solar system?

Chapter 10

The Cosmic Microwave Background

The time for discussion of the global dynamics of the Universe is over. We now move on to the question of why it is said to be the **Hot Big Bang**. From now on I will concentrate on the case of a flat Universe with no cosmological constant; this sounds significant but we will see during Chapter 13 that this is always a good approximation during the early evolution of the Universe, even if the present Universe is not flat or possesses a cosmological constant.

10.1 Properties of the microwave background

The crucial observation which swayed the Big Bang/Steady State Universe debate in favour of the former was the detection of the cosmic microwave background radiation reported in 1965. This radiation bathes the Earth from all directions, and is now known to accurately take on the form of a black-body with temperature

$$T_0 = 2.725 \pm 0.001 \text{ Kelvin}, \tag{10.1}$$

as shown in Figure 2.4. As a first step, let's work out how much energy that corresponds to, in comparison to the critical density.

We studied the properties of thermal distributions in Section 2.5.2. The black-body spectrum is given by equation (2.8). We found the total energy density $\epsilon_{\rm rad}$ of radiation at temperature T by integrating the energy density over the black-body distribution, obtaining equation (2.10)

$$\epsilon_{\rm rad} \equiv \rho_{\rm rad} c^2 = \alpha T^4 , \tag{10.2}$$

where

$$\alpha \equiv \frac{\pi^2 k_{\rm B}^4}{15 \hbar^3 c^3} = 7.565 \times 10^{-16}\,{\rm J\,m^{-3}\,K^{-4}} , \tag{10.3}$$

is the radiation (or black-body) constant. Evaluating equation (10.2) for the observed temperature gives the present energy density of radiation

$$\epsilon_{\mathrm{rad}}(t_0) = 4.17 \times 10^{-14}\,\mathrm{J\,m^{-3}}\,. \tag{10.4}$$

Writing this in terms of the critical density, equation (6.5), remembering to convert energy density to mass density by dividing by c^2, yields

$$\Omega_{\mathrm{rad}} = 2.47 \times 10^{-5} h^{-2}\,. \tag{10.5}$$

So the radiation in the microwave background (which in fact dominates the energy density in radiation of all wavelengths) is a small but not completely negligible fraction of the critical density. However, it is quite a lot smaller than the density which we presently see in stars etc, even without further considering that most of the matter in the Universe is probably dark matter.

However, we know how the density of radiation evolves with the expansion of the Universe, equation (5.18):

$$\rho_{\mathrm{rad}} \propto \frac{1}{a^4}\,. \tag{10.6}$$

Combined with equation (10.2), this implies the following crucial equation

$$T \propto \frac{1}{a}\,. \tag{10.7}$$

This means that the Universe cools as it expands. Since today it has a temperature of about 3K, that means that at earlier times it must have been much hotter. In fact, since the further back in the past we consider, the smaller the Universe was, it must have been arbitrarily hot in its earliest stages.

If the temperature is changing as the Universe evolves, then the thermal distribution must evolve too. However, the energy density distribution of the black-body distribution

$$\epsilon(f)df = \frac{8\pi h}{c^3}\,\frac{f^3 df}{\exp\left(hf/k_{\mathrm{B}}T\right) - 1}\,, \tag{10.8}$$

that we saw in Section 2.5.2 [equation (2.8)] has a special property, shown in Figure 10.1. As the Universe expands, the frequency f reduces in proportion to $1/a$, but the black-body form is preserved at a lower temperature $T_{\mathrm{final}} = T_{\mathrm{initial}} \times a_{\mathrm{initial}}/a_{\mathrm{final}}$. This works for two reasons. The first is because the denominator is only a function of f/T and not f and T separately, and so the reduction of f can be absorbed by an equivalent reduction in T. The second is because the f^3 on the numerator scales as the inverse volume, corresponding to the evolution of the photon number density as the Universe expands. So as the Universe expands and cools, the photon distribution continues to correspond to a thermal distribution, but one with ever lower temperature. Consequently, as long as at some early stage interactions were frequent enough to establish a thermal distribution, it will persevere even if at a later stage particle interactions become infrequent.

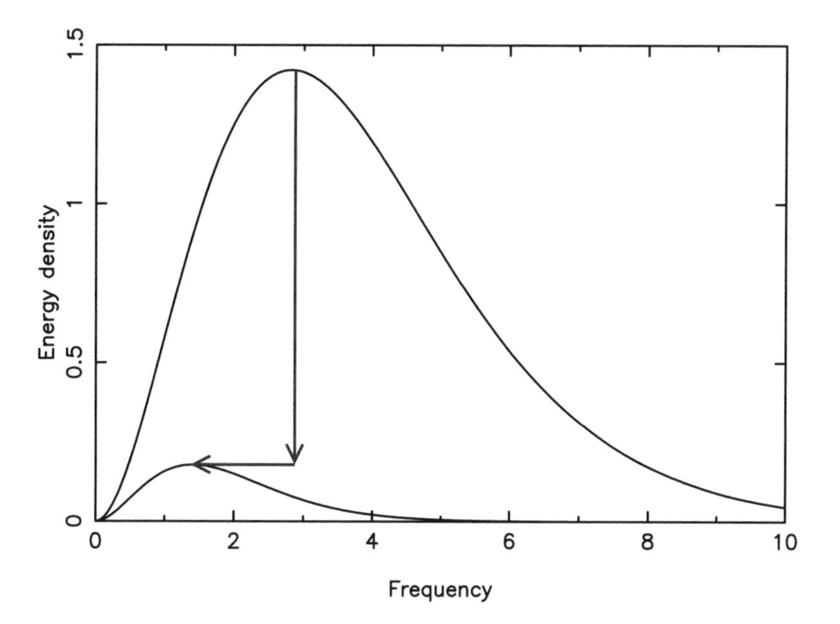

Figure 10.1 The evolution of the black-body spectrum as the Universe expands. The expansion reduces the number density of photons, while the redshifting reduces their frequency. In combination, these two effects map the spectrum onto a new black-body at a lower temperature. [The axes match Figure 2.8, using the higher temperature for each curve. The cooling is by a factor two.]

10.2 The photon to baryon ratio

Before proceeding to the origin of the microwave background, let's reconsider it in terms of the numbers of particles. You'll recall, from Section 5.4, that since particles cannot simply disappear, then so long as interactions are negligible particle number densities simply reduce in inverse proportion to the volume, $n \propto 1/a^3$. This is true of both protons and neutrons, collectively known as baryons, and of the photons making up the microwave background. The ratio of the number of photons to the number of baryons is therefore a constant, preserved as the Universe expands. How many photons are there per baryon?

We've just seen that the present energy in the microwave background is

$$\epsilon_{\rm rad}(t_0) = 4.17 \times 10^{-14}\,{\rm J\,m^{-3}}. \tag{10.9}$$

The typical energy of a photon of light in a thermal distribution is (see page 15)

$$E_{\rm mean} \simeq 3k_{\rm B}T = 7.05 \times 10^{-4}\,{\rm eV}\,, \tag{10.10}$$

for a temperature $T = 2.725\,{\rm K}$. Converting electron-volts to Joules (page xiv) and dividing the energy density by the mean energy we find the present number density of photons is

$$n_\gamma = 3.7 \times 10^8 \, \mathrm{m}^{-3} \,. \tag{10.11}$$

There are nearly a billion microwave background photons in every cubic metre!

Now we need to compare this to the number density of baryons. The best observations relate to nucleosynthesis, which is the topic of Chapter 12; however, we already saw the result quoted in Section 9.1.2. The density parameter in baryons is

$$\Omega_\mathrm{B} \simeq 0.02 \, h^{-2} \,. \tag{10.12}$$

We convert this into an energy density using the critical density, obtaining

$$\epsilon_\mathrm{B} = \rho_\mathrm{B} c^2 = \Omega_\mathrm{B} \rho_\mathrm{c} c^2 \simeq 3.38 \times 10^{-11} \, \mathrm{J \, m}^{-3}. \tag{10.13}$$

The baryon energy density is about a thousand times larger than the density parameter in radiation, equation (10.5), but the individual baryons have much more energy, the proton and neutron rest masses being about 939 MeV. We find

$$n_\mathrm{B} = 0.22 \, \mathrm{m}^{-3} \,. \tag{10.14}$$

Although the total energy density in baryon considerably exceeds that in radiation, there are vastly more photons than there are baryons. In fact, there are about 1.7×10^9 photons for every baryon.

10.3 The origin of the microwave background

We are now in a position to discuss the origin of the microwave background. The crucial ingredient we need is that a hydrogen atom has a minimum ionization energy; if an electron finds its way into the ground state then 13.6 eV of energy is needed to free it. At the very least, 10.2 eV is needed to raise it to its first excited state, from which a further 3.4 eV will ionize it. As long as the Universe is hot enough, photons will easily have this energy and are able to keep the hydrogen fully ionized.

Let's begin by considering a suitably early time, say when the Universe was one millionth of its present size. At that time the temperature would have been about 3 000 000 K. Such a temperature was high enough that the typical energy of a photon in the thermal distribution was considerably more than the ionization energy of hydrogen atoms, so atoms would not have been able to exist at that epoch; any electron trying to bind to a proton would immediately be blasted away again by collision with a photon of light. The Universe at that time was therefore a sea of free nuclei and electrons, and because photons interact strongly with free electrons (via Thomson scattering), the mean free path of any photon was short (approximately $1/n_e\sigma_e$ where n_e is the electron number density and σ_e the Thomson scattering cross-section). So we picture a sea of frequently-colliding particles, forming an ionized plasma. This situation is actually not very exotic; if you calculate the density of material at that time you will find it's very low — considerably less than water — and it's very easy to heat a gas up until it becomes a plasma. The physics is all extremely well tested and understood.

As the Universe expanded and cooled, the photons of light lost energy and became less and less able to ionize any atoms that form. The situation is exactly that of the photoelectric effect, where long-wavelength photons, however numerous, are unable to knock electrons out of metal atoms. Eventually all the electrons found their way into the ground state and the photons were no longer able to interact at all. Over a short interval of time, the Universe suddenly switched from being opaque to being completely transparent. The photons were then able to travel unimpeded for the entire remainder of the Universe's evolution. This process is known as **decoupling**.

The simplest estimate of when the microwave background formed comes from equating the mean photon energy at a given temperature to the ionization energy. The mean energy of a photon in a black-body distribution at temperature T is $E \simeq 3k_{\rm B}T$, as we saw in Section 2.5.2. Since $k_{\rm B} = 8.62 \times 10^{-5}\,{\rm eV\,K^{-1}}$, the temperature at formation of the microwave background would be

$$T \simeq \frac{13.6\,{\rm eV}}{3k_{\rm B}} = 50\,000\,{\rm K}\,, \tag{10.15}$$

if this procedure were valid. In fact, this estimate is not quite sophisticated enough, because we have yet to account for our discovery, in the previous section, that there are far more photons in the Universe than electrons, about a factor of 10^9 more. Because of this, even when the mean photon energy dropped below 13.6 eV, there were still high-energy photons in the tail of the distribution able to ionize any atoms that formed — see equation (2.7) and Figure 2.7 on page 13.

An accurate calculation of the temperature of decoupling requires a lot of physics, some of which I will outline in an optional section following this one. However we can at least estimate the order of the effect simply using the Boltzmann suppression factor, making the assumption that we only need something like one ionizing photon per atom to keep the Universe ionized. [Note that the high-energy photons keep the electrons out of atoms, and then all the remaining photons can interact with the free electrons thus created.] In its crudest form, the Boltzmann suppression says that the fraction of photons with energy exceeding I is given approximately by $\exp(-I/k_{\rm B}T)$, leading to the expression

$$T_{\rm dec} = \frac{13.6\,{\rm eV}}{k_{\rm B}\,\ln(1.7 \times 10^9)} \simeq 7400\,{\rm K}\,. \tag{10.16}$$

We can do a little better by actually integrating over the photon distribution function, equation (2.7), which indicates a significant prefactor to the Boltzmann suppression, reducing the estimate to $5700\,{\rm K}$ (see Problem 10.5). This is in fact fairly close to the right answer, which is that decoupling occurred when the Universe was at a temperature of about 3000 K. This temperature is known as the decoupling temperature.

Comparing this to the present temperature, we conclude, using equation (10.7), that decoupling happened when the Universe was about one-thousandth of its present size, with $a_{\rm dec} \simeq 1/1000$ assuming we have normalized $a(t_0) = 1$.

Thus, the reason why the microwave background is so accurately given by a thermal distribution is that it was once in a highly-interacting thermal state when the Universe was much hotter. As we've seen, the black-body form is preserved as the Universe expands

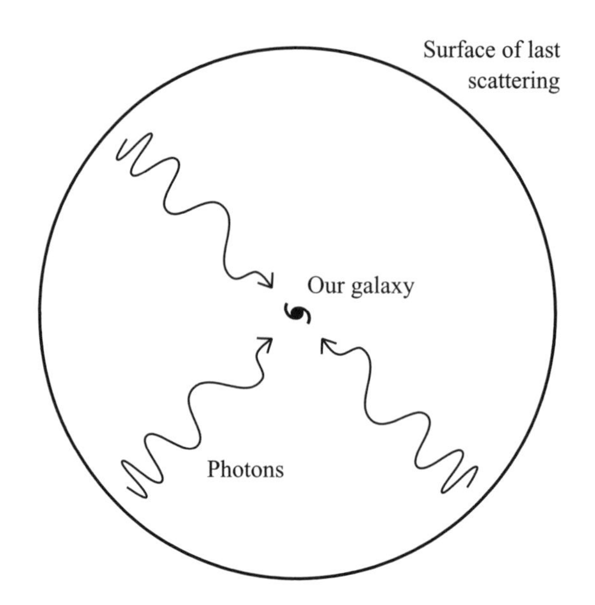

Figure 10.2 Because they have travelled towards us uninterrupted since the Universe was at 3000 K, the photons making up our microwave background originated on the surface of a sphere, known as the surface of last scattering, a considerable distance away from our own galaxy. If observers exist in other galaxies, they will see microwaves coming from the surface of a different sphere centred around their location. At the last-scattering surface, the photons had a much higher frequency, which has been redshifted as the photons travel towards us.

and cools. The Hot Big Bang theory therefore gives a simple explanation of this crucial observation. In the Steady State theory, all radiation is supposed to originate in stars and so is at high frequency and is not a perfect black-body; one has to resort to a thermalizing mechanism such as whiskers of iron, which somehow managed to thermalize this into low-energy radiation in the recent past without preventing us from seeing distant objects. It has never been satisfactorily demonstrated that this can be achieved even allowing the *ad hoc* assumptions that the Steady State scenario requires.

Since decoupling happened when the Universe was only about one thousandth of its present size, and the photons have been travelling uninterrupted since then, they come from a considerable distance away. Indeed, a distance close to the size of the observable Universe. Those we see originate on the surface of a very large sphere centred on our location, shown in Figure 10.2, called the **surface of last scattering**. Its radius is of order $6000\,h^{-1}$ Mpc (see Problem 10.5). Of course, there is nothing special about this particular surface, except that it happens to be at just the right distance that the photons have reached us by today. There are photons originating at every point, and observers in different parts of the Universe (if there are any!) will see photons originating from different large spheres, of the same radius, centred on their location.

When the photons set out, their temperature was about 3000 K and their frequencies were much higher than now, so that they weren't actually microwaves at that time. We'll see in the next chapter that the age of the Universe at that time was about 350 000 yrs. As the photons travel, the Universe expands and they cool, maintaining their thermal form, until they are detected on Earth at a temperature a little below 3K. By this time, redshifting has placed them in the microwave region of the electromagnetic spectrum.

10.4 The origin of the microwave background (optional advanced treatment)

The derivation of the temperature when the cosmic microwave background formed given in the last section is not entirely satisfactory, and in this optional section I give a brief description of a more accurate calculation, based loosely on the textbook of Kolb & Turner. First of all, we should acknowledge that there are two separate processes going on, which I have up until now taken as one and the same. **Recombination** refers to the epoch where electrons joined the nuclei to create atoms ('recombination' is however a misnomer in that the electrons and nuclei were never previously combined). **Decoupling** refers to the epoch after which the photons will not scatter again. If recombination were instantaneous and complete, the two would coincide, but in practice each process takes some time and decoupling follows recombination. An improved calculation of the recombination epoch uses the Saha equation, which computes the ionization fraction of a gas in thermal equilibrium.

The Saha equation is derived by assuming that only hydrogen is present, and considers the distribution functions for hydrogen, free protons and electrons, which are assumed to be in thermal and chemical equilibrium (a chemical potential has to be included to enforce baryon number conservation). Defining the ionization fraction as $X \equiv n_{\rm p}/n_{\rm B}$, where $n_{\rm p}$ and $n_{\rm B}$ are the number densities of free protons and of baryons, the equilibrium abundance can be shown to be

$$\frac{1-X}{X^2} \simeq 3.8 \, \frac{n_{\rm B}}{n_\gamma} \left(\frac{k_{\rm B}T}{m_{\rm e}c^2} \right)^{3/2} \exp\left(\frac{13.6\,{\rm eV}}{k_{\rm B}T} \right) . \tag{10.17}$$

This complicated formula properly includes the distribution functions, and also allows for the large ratio of photons to baryons, $n_{\rm B}/n_\gamma \simeq 6 \times 10^{-10}$.

If the right-hand side of the Saha equation is small, then X will be close to one corresponding to full ionization. For example, this will be true if the temperature is much greater than the binding energy (while still low enough to avoid spontaneous electron–positron creation), due to the small baryon-to-photon ratio. As the temperature cools, the right-hand side will become large and the ionization ratio will fall towards zero. However both the factors in front of the exponential are small numbers, and these are overcome by the exponential only once the temperature is well below the binding energy. Note that as well as the baryon-to-photon ratio discussed earlier, there is a significant prefactor $\left(k_{\rm B}T/m_{\rm e}c^2\right)^{3/2}$ coming from correct treatment of the distribution functions.

A standard definition of recombination is $X_{\rm rec} = 0.1$, corresponding to the process being 90 percent complete. The Saha equation has to be solved iteratively or numerically, once the precise value of $n_{\rm B}/n_\gamma$ has been chosen, and this gives $k_{\rm B}T_{\rm rec} \simeq 0.31\,{\rm eV}$, implying $T_{\rm rec} \simeq 3600\,{\rm K}$.

In fact not all electrons manage to combine with atoms, because eventually the remaining electrons are so rare that they are unable to find their corresponding nuclei. However the Saha equation cannot be used to predict the residual ionization as equilibrium will have broken down by then, and so a yet more sophisticated calculation is needed. The residual ionization is expected to be of order 10^{-3}.

Having found the ionization history of the Universe, the epoch of decoupling can be formally defined as the epoch when the duration of the photon mean free path equals the age of the Universe. The mean free path by this stage grows much more rapidly than the size of the observable Universe, so the photons are then unlikely to interact at any subsequent epoch. The result has a very weak dependence on the matter and baryon densities, and can be shown to be $T_{\rm dec} \simeq 3000\,{\rm K}$ as given in the previous section.

Problems

10.1. If the microwave background has a temperature of about 3K, why does a microwave oven heat food up rather than cool it down? Also, if microwaves can't interact with atoms as they have insufficient energy to shift the electrons up energy levels, how can a microwave oven heat food?

10.2. The energy density ϵ in radiation is related to its temperature by

$$\epsilon_{\rm rad} = \rho_{\rm rad} c^2 = \alpha T^4 \,.$$

Compute the temperature when the Universe was one second old, using the Friedmann equation and the radiation-dominated solution $a(t) \propto t^{1/2}$. [You'll need some of the constants listed on page xiv.]

What was the corresponding mass density at that time? Compare it with that of water. How old is the Universe when its density matches that of water?

10.3. Suppose we live in a closed Universe ($k > 0$), which will recollapse some time in the future. What will the temperature be when the Universe has gone through its maximum size and then shrunk back to its present size?

10.4. The present number density of electrons in the Universe is the same as that of protons, about $0.2\,{\rm m}^{-3}$. Consider a time long before the formation of the microwave background, when the scale factor was one millionth of its present value. What was the number density of electrons then? Given that the electron mass–energy is 0.511 MeV, do you expect electrons to be relativistic or non-relativistic at that time?

The cross-section for the scattering of photons off electrons is the Thomson cross-section $\sigma_{\rm e} = 6.7 \times 10^{-29} {\rm m}^2$. Given that the mean free path (i.e. the typical distance travelled between interactions) of photons through an electron gas of number density $n_{\rm e}$ is $d \simeq 1/n_{\rm e}\sigma_{\rm e}$, compute the mean free path for photons when the scale factor was one millionth its present value.

From the mean free path, calculate the typical time between interactions, the speed of light being $3 \times 10^8\,{\rm m\,sec}^{-1}$. Compare the interaction time with the age of the Universe at that time, which would be about 10 000 years. What is the significance of the comparison?

10.5. Integration of the Planck function (which you can try yourself if you have time on your hands) shows that if $I \gg k_{\mathrm{B}}T$ the fraction of photons of energy greater than I is

$$\frac{n(>I)}{n} \simeq \left(\frac{I}{k_{\mathrm{B}}T}\right)^2 \exp\left(-\frac{I}{k_{\mathrm{B}}T}\right) .$$

Either numerically or by iteration, find the temperature such that there is one ionizing photon per baryon.

10.6. Use the age of the Universe to estimate the radius of the last-scattering surface, assuming critical density. Why might this underestimate the true value? Assuming a typical galaxy has mass $10^{11} M_{\odot}$, and using the critical density given in equation (6.6), estimate the number of galaxies in the observable Universe. How many protons are there in the observable Universe?

Chapter 11

The Early Universe

Now that we understand the behaviour of the radiation, we can consider the entire thermal history of the Universe. The best approach is to start from the present and work backwards, and see how far our understanding can take us.

At the present we have some idea of the constituents of the Universe, at least up to the uncertainty in cosmological parameters such as h. The relativistic particles come in two varieties, photons and neutrinos. The photon density we have already found to be $\Omega_{\rm rad} = 2.47 \times 10^{-5}\, h^{-2}$, equation (10.5). The neutrinos present more of a challenge, because neutrinos are fiendishly hard to detect. For example, to detect the neutrinos even from something as optically bright as our own Sun requires delicate underground experiments involving huge tanks of material. Direct detection of a thermal cosmological neutrino background is presently orders of magnitude beyond our technical expertise. To estimate the properties of the neutrino background, we must for now resort to purely theoretical arguments.

Within this main body of this book, I will make the common assumption that as far as cosmology is concerned the neutrinos can be treated as massless particles. There is in fact now substantial experimental evidence that neutrinos have some mass, though it is unclear whether this is large enough to have cosmological effects, and Advanced Topic 3 will study the effects of neutrino mass in some detail. Under the massless assumption, theoretical calculations of the present neutrino density give a famous and bizarre-looking result

$$\Omega_\nu = 3 \times \frac{7}{8} \times \left(\frac{4}{11}\right)^{4/3} \Omega_{\rm rad} = 0.68\, \Omega_{\rm rad} = 1.68 \times 10^{-5} h^{-2}\,, \qquad (11.1)$$

the steps to which you can follow in Problem 11.1 and in Advanced Topic 3. The amount of energy expected in the cosmic neutrino background is similar to that in the cosmic microwave background.

Adding together the photon and neutrino densities gives the complete matter density in relativistic particles

$$\Omega_{\rm rel} = 4.15 \times 10^{-5} h^{-2}\,. \qquad (11.2)$$

Since this is well below the observed density of the matter in the Universe, most of the matter in the present Universe is non-relativistic. The density of non-relativistic material is simply Ω_0, which is expected to be around 0.3.

We know the dependences of both relativistic and non-relativistic matter densities on the expansion, reducing as $1/a^4$ and $1/a^3$ respectively. Their ratio, expressed using the density parameter, therefore behaves as

$$\frac{\Omega_{\rm rel}}{\Omega_{\rm mat}} = \frac{4.15 \times 10^{-5}}{\Omega_0 h^2} \frac{1}{a} , \tag{11.3}$$

where the constant of proportionality has been fixed by the present values and it is assumed we normalize $a(t_0) = 1$. With this, we can compute the relative amounts of relativistic and non-relativistic material for any given size of the Universe. For example, at decoupling we found $a_{\rm dec} \simeq 1/1000$, so the ratio at decoupling is given by

$$\frac{\Omega_{\rm rel}}{\Omega_{\rm mat}} = \frac{0.04}{\Omega_0 h^2} . \tag{11.4}$$

Unless the combination $\Omega_0 h^2$ is very small, there will be more non-relativistic matter than not at the time of decoupling; the Universe is said to be matter dominated.

However, considering earlier times that state of affairs cannot persevere for long; when

$$a = a_{\rm eq} = \frac{1}{24\,000\,\Omega_0 h^2} , \tag{11.5}$$

the densities of matter and radiation would be the same. This is known as the epoch of **matter–radiation equality**. At all earlier times, the relativistic particles would dominate the Universe.

We now have enough information to calculate the full temperature versus time relationship for the Universe, assuming an instantaneous transition between radiation domination and matter domination. Since $T \propto 1/a$, and we know how a behaves in each of those regimes, we can immediately write down the appropriate results.

An acceptable approximation is to set $k = 0$ and $\Lambda = 0$, as even if they are present now they would be negligible early on. Then the scale factor grows as $a \propto t^{2/3}$, giving the relation $T \propto t^{-2/3}$. Fixing the proportionality constant assuming the Universe is presently 12 billion years old (a slight underestimate to compensate for ignoring Λ) gives

$$\frac{T}{2.725\,{\rm K}} = \left(\frac{4 \times 10^{17}\,{\rm sec}}{t}\right)^{2/3} . \tag{11.6}$$

This holds for

$$T < T_{\rm eq} = \frac{2.725\,{\rm K}}{a_{\rm eq}} = 66\,000\,\Omega_0 h^2\,{\rm K} . \tag{11.7}$$

The time of matter–radiation equality is then given by

$$t_{\rm eq} \simeq 1.0 \times 10^{11}\,\Omega_0^{-3/2} h^{-3}\,{\rm sec} \simeq 3400\,\Omega_0^{-3/2} h^{-3}\,{\rm yrs} . \tag{11.8}$$

As decoupling happened after matter–radiation equality, we can apply equation (11.6) with $T_{\mathrm{dec}} \simeq 3000\,\mathrm{K}$ to find the age of the Universe at decoupling

$$t_{\mathrm{dec}} \simeq 10^{13}\,\mathrm{sec} = 350\,000\,\mathrm{yrs}\,. \tag{11.9}$$

At temperatures above T_{eq}, radiation domination takes over and, from the expansion law $a \propto t^{1/2}$, the temperature–time relation becomes

$$\frac{T}{T_{\mathrm{eq}}} = \left(\frac{t_{\mathrm{eq}}}{t}\right)^{1/2}\,, \tag{11.10}$$

where the constant of proportionality is fixed by the values at matter–radiation equality.

However, during radiation domination we can obtain a more accurate result directly from the Friedmann equation. Write

$$H^2 = \frac{8\pi G}{3}\,\rho = \frac{8\pi G}{3} \times 1.68 \times \frac{\alpha T^4}{c^2}\,, \tag{11.11}$$

and substitute in for all the constants, remembering that radiation domination gives $a \propto t^{1/2}$ and hence $H = 1/2t$. The factor 1.68 allows for the neutrinos. This then gives

$$\left(\frac{1\,\mathrm{sec}}{t}\right)^{1/2} \simeq \frac{T}{1.3 \times 10^{10}\,\mathrm{K}} = \frac{k_{\mathrm{B}} T}{1.1\,\mathrm{MeV}}\,. \tag{11.12}$$

This means that when the Universe was one second old, the temperature would have been about 1.3×10^{10} Kelvin and the typical particle energy about 1.1 MeV.

The temperature–time relation for the Universe is illustrated in Figure 11.1.

Knowing the typical energy of the radiation as a function of time allows us to construct a history of interesting eras in the evolution of the Universe. Let's begin at the present and consider running time backwards, so that the Universe gets hotter as we go to earlier and earlier times.

We've already discussed decoupling, which was when the microwave background formed. It corresponds to the last time photons were energetic enough to knock electrons out of atoms, at a temperature of about 3000 K. Looking at equation (11.7), we see that decoupling almost certainly happened during the matter-dominated era. However, continuing to run time backwards, we learn that a little earlier the radiation would have been the dominant constituent of the Universe, according to equation (11.5). The transition occurred at a temperature $T_{\mathrm{eq}} = 66\,000\,\Omega_0 h^2\,\mathrm{K}$.

As we contemplate earlier times, the Universe was ever hotter, but we have to consider quite early times before that extra energy has a significant effect. At times early enough that the temperature exceeded 10^{10} Kelvin, the typical photon energies were comparable to nuclear binding energies, which are of order an MeV; this would have been the case when the Universe was around one second old. When the Universe was younger than this, the photons were energetic enough to destroy nuclei, by splitting protons and neutrons away from each other. So at any time before an age of one second the Universe would have been a sea of separate protons, neutrons, electrons etc, strongly interacting with each

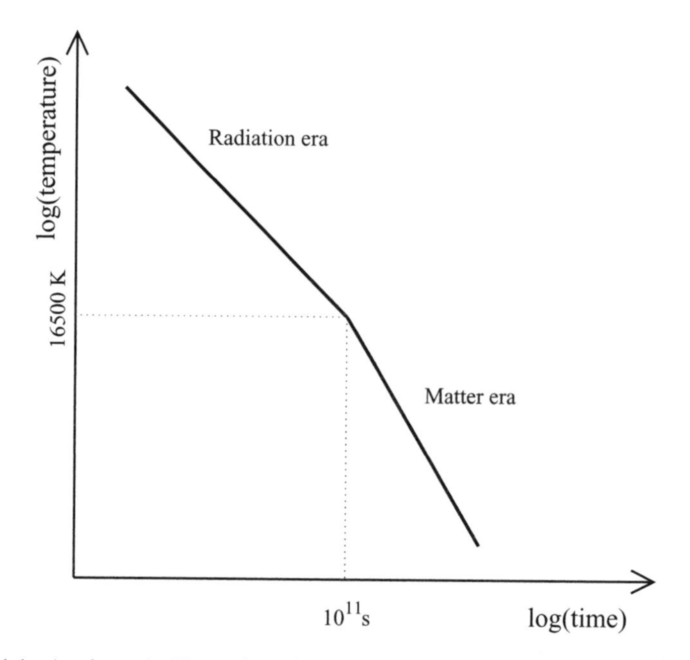

Figure 11.1 A schematic illustration of the temperature–time relation, assuming $\Omega_0 = 1$ and $h = 0.5$. When the radiation era ends the expansion rate increases and the temperature cools more quickly.

other. The transition from free protons and neutrons into atomic nuclei is the topic of the next chapter.

Going back further in time, when the temperature was even hotter, the situation becomes less clear, because the typical energies start to be so high that the laws of physics become less well known. It is believed that at 10^{12} Kelvin it stops making sense even to think of protons and neutrons; instead their constituent quarks are free to wander around in a dense sea (rather reminiscently of the way that in some molecules the electrons are not associated with any particular nucleus). The transition where quarks first condense into protons and neutrons is known as the quark–hadron phase transition [hadron being the technical term for bound states of quarks, either baryons (three quarks) or mesons (a quark and an anti-quark)]. Theoretically this picture is appealing, but observational evidence, obtained by colliding heavy nuclei together, is so far scanty at best.

The highest particle energies that have been achieved on Earth are generated by particle accelerators and are around 100 GeV (where GeV is a giga-electron volt, i.e. one thousand MeV), corresponding to an effective temperature of about 10^{15} Kelvin. This is the highest energy at which we have direct evidence of the physical behaviour of fundamental particles, and that temperature was achieved only 10^{-10} seconds after the Big Bang itself. Earlier yet lies the realm of the very early Universe, where speculations concerning laws of physics such as the unification of fundamental forces must be used. A variety of possible behaviours have been proposed; one particularly prominent idea is cosmological inflation, which I'll come to shortly.

The different eras are summarized in Table 11.1. Note that I haven't mentioned the

Table 11.1 Different stages of the Universe's evolution (taking $\Omega_0 = 0.3$ and $h = 0.72$). Some numbers are approximate.

Time	Temperature	What's going on?
$t < 10^{-10}$ s	$T > 10^{15}$ K	Open to speculation!
10^{-10} s $< t < 10^{-4}$ s	10^{15} K $> T > 10^{12}$ K	Free electrons, quarks, photons, neutrinos; everything is strongly interacting with everything else.
10^{-4} s $< t <$ 1 s	10^{12} K $> T > 10^{10}$ K	Free electrons, protons, neutrons, photons, neutrinos; everything is strongly interacting with everything else.
1 s $< t < 10^{12}$ s	10^{10} K $> T >$ 10000 K	Protons and neutrons have joined to form atomic nuclei, and so we have free electrons, atomic nuclei, photons, neutrinos; everything is strongly interacting with everything else except the neutrinos, whose interactions are now too weak. The Universe is still radiation dominated.
10^{12} s $< t < 10^{13}$ s	10000 K $> T >$ 3000 K	As before, except that now the Universe is matter dominated.
10^{13} s $< t < t_0$	3000 K $> T >$ 3 K	Atoms have now formed from the nuclei and the electrons. The photons are no longer interacting with them, and are cooling to form what we will see as the microwave background.

dark matter, since so little is known about it, but it is most likely present at all these epochs and, at least at the later stages, cannot have significant interactions with anything else or it wouldn't be dark.

Problems

11.1. This question indicates the path to the neutrino density Ω_ν of equation (11.1). Remember that there are three different families of neutrino, each contributing to the density. Very early on the Universe was so hot and dense that even neutrinos would interact sufficiently to become thermalized.

The neutrino temperature is predicted to be lower than the photon temperature, the reason being electron–positron annihilations which feed energy into the photon energy density but not the neutrino one. This boosts the photon temperature relative to the neutrinos by a factor $\sqrt[3]{11/4}$. Compute $\Omega_\nu/\Omega_{\rm rad}$, assuming at this stage that the radiation constant is the same in each case.

In fact, the equivalent of the radiation constant for neutrinos is lower than that for photons, by a factor $7/8$. [The fundamental reason for this is that neutrinos obey Fermi–Dirac statistics rather than Bose–Einstein ones as photons do; their equivalent of equation (2.7) has $+1$ rather than -1 on the denominator.] Correct your estimate of $\Omega_\nu/\Omega_{\rm rad}$ to include this.

11.2. In Section 10.2, we learned that the number density of photons in the microwave background is $n_\gamma \simeq 3.7 \times 10^8\,{\rm m}^{-3}$. Assuming neutrinos are massless, estimate the neutrino number density. Estimate how many cosmic neutrinos pass through your body each second.

11.3. The temperature at the core of the Sun is around 10^7 K. How old was the Universe when it was this hot? Was it matter dominated or radiation dominated at that time?

At the CERN collider, typical particle energies are of order of 100 GeV. How old was the Universe when typical particle energies were around this size? What was the temperature at this time?

11.4. Estimate $\Omega_{\rm rad}$ at the time of decoupling, stating clearly any assumptions.

Chapter 12

Nucleosynthesis: The Origin of the Light Elements

The abundance of elements in the Universe provides the final, and in many ways most compelling, piece of evidence supporting the Hot Big Bang theory. Historically it was assumed that the first stars began their life made from hydrogen, with heavier elements being generated via nuclear fusion reactions as they burned (later generations of stars formed from gas that contained heavier elements produced by the first stars). While this is certainly the process giving rise to the heavy elements, it was eventually recognized that all the light elements — deuterium, helium-3, lithium and especially helium-4 — could not have been created in this manner. Instead, as one looks to younger and younger stars, these approach non-zero abundances, which the stars seem to begin their lives with. These abundances are apparently those of the primordial gas from which the stars formed, and the question is whether or not they can be explained by the Hot Big Bang theory.

The processes which give rise to nuclei parallel those which we have already examined for atoms in Chapter 10. A typical nuclear binding energy is around 1 MeV, and so if typical photon energies exceed this, then nuclei will be immediately dissociated. This energy is about 100 000 times greater than the electron binding energy, and so the corresponding temperature is higher by this factor. The formation of nuclei in the Universe therefore took place at a much earlier stage in the Universe's history; from the temperature–time relation of the last chapter, equation (11.12), we see that this should have happened when the Universe was about one second old. The process is known as **nucleosynthesis**.

12.1 Hydrogen and Helium

I'll give a simplified analysis, which assumes that only helium-4, the most stable of the light nuclei, was formed, with the leftover material remaining as hydrogen nuclei (i.e. individual protons). Three pieces of physics are important:

- Protons are lighter than neutrons ($m_{\rm p}\, c^2 = 938.3$ MeV; $m_{\rm n}\, c^2 = 939.6$ MeV).
- Free neutrons don't survive indefinitely, but instead decay into protons with a surprisingly long half-life of $t_{\rm half} = 614$ sec.

- There exist stable isotopes of light elements, and neutrons bound into them do not decay.

At high temperatures the Universe contains protons and neutrons in thermal equilibrium at high energies. As it cools, these at some point stop being free particles and are able to bind into nuclei.

We start our discussion at a time before the nuclei form, but late enough that the temperature is sufficiently low that the protons and neutrons are non-relativistic, meaning $k_{\mathrm{B}}T \ll m_{\mathrm{p}}c^2$. When this is satisfied the particles will be in thermal equilibrium and satisfy a Maxwell–Boltzmann distribution, in which the number density N is given by

$$N \propto m^{3/2} \exp\left(-\frac{mc^2}{k_{\mathrm{B}}T}\right). \tag{12.1}$$

[I'm using N for number density in this section to avoid confusion with 'n' for neutron; N is the same as the number density n of Section 5.4 and elsewhere.] The constant of proportionality is the same for each particle species and isn't needed. The relative densities of neutrons and protons will be

$$\frac{N_{\mathrm{n}}}{N_{\mathrm{p}}} = \left(\frac{m_{\mathrm{n}}}{m_{\mathrm{p}}}\right)^{3/2} \exp\left[-\frac{(m_{\mathrm{n}} - m_{\mathrm{p}})\,c^2}{k_{\mathrm{B}}T}\right]. \tag{12.2}$$

The prefactor is always very close to one as the particle masses are so similar. The exponential factor is also close to one as long as the temperature exceeds the proton–neutron mass difference of 1.3 MeV, so while $k_{\mathrm{B}}T \gg (m_{\mathrm{n}} - m_{\mathrm{p}})c^2$ the numbers of protons and of neutrons in the Universe will be almost identical.

The reactions converting neutrons to protons and vice versa are

$$\mathrm{n} + \nu_{\mathrm{e}} \longleftrightarrow \mathrm{p} + \mathrm{e}^- \tag{12.3}$$

$$\mathrm{n} + \mathrm{e}^+ \longleftrightarrow \mathrm{p} + \bar{\nu}_{\mathrm{e}} \tag{12.4}$$

where ν_{e} is an electron neutrino and $\bar{\nu}_{\mathrm{e}}$ its anti-particle. As long as these interactions proceed sufficiently rapidly, the neutrons and protons will remain in thermal equilibrium with abundance determined by equation (12.2). A calculation of the interaction rate is beyond the scope of this book, but indicates that reactions proceed quickly until the temperature reaches $k_{\mathrm{B}}T \simeq 0.8$ MeV, after which the rate becomes much longer than the age of the Universe. At that temperature, the relative abundances of protons and neutrons become fixed. As this temperature is slightly less than the neutron–proton mass-energy difference, the exponential in equation (12.2) has become important and the relative number densities are

$$\frac{N_{\mathrm{n}}}{N_{\mathrm{p}}} \simeq \exp\left(-\frac{1.3\ \mathrm{MeV}}{0.8\ \mathrm{MeV}}\right) \simeq \frac{1}{5}. \tag{12.5}$$

From this time onwards, the only process which can change the abundances is the decay of free neutrons.

The production of light elements then has to go through a complex reaction chain, with nuclear fusion forming nuclei and the high-energy tail of the photon distribution breaking them up again (just as at the formation of the microwave background). The sort of reactions which are important (but far from a complete set) are

$$\mathrm{p} + \mathrm{n} \;\rightarrow\; \mathrm{D}\,; \tag{12.6}$$

$$\mathrm{D} + \mathrm{p} \;\rightarrow\; {}^{3}\mathrm{He}\,; \tag{12.7}$$

$$\mathrm{D} + \mathrm{D} \;\rightarrow\; {}^{4}\mathrm{He}\,, \tag{12.8}$$

where 'D' stands for a deuterium nucleus and 'He' a helium one. The destruction processes happen in the opposite direction; they become less and less important as the Universe cools and eventually the build-up of nuclei can properly proceed. It turns out that this happens at an energy of about 0.06 MeV. I won't attempt a derivation of that number, though I note that it can be estimated by a similar 'high-energy tail' argument to that of Chapter 10, this time applied to the deuterium binding energy of 2.2 MeV. Once the neutrons manage to form nuclei, they become stable.

The delay until 0.06 MeV before nuclei such as helium-4 appear is long enough that the decay of neutrons into protons is not completely negligible, though most of the neutrons do survive. To figure out how many neutrons decay, we need to know how old the Universe is at a temperature $k_{\mathrm{B}}T \simeq 0.06\,\mathrm{MeV}$. We found this in the last chapter, equation (11.12); the age is $t_{\mathrm{nuc}} \simeq 340\,\mathrm{s}$, surprisingly close to the neutron half-life of $t_{\mathrm{half}} = 614\,\mathrm{s}$. The neutron decays reduce the neutron number density by $\exp(-\ln 2 \times t_{\mathrm{nuc}}/t_{\mathrm{half}})$ giving

$$\frac{N_{\mathrm{n}}}{N_{\mathrm{p}}} \simeq \frac{1}{5} \times \exp\left(-\frac{340\,\mathrm{s} \times \ln 2}{614\,\mathrm{s}}\right) \simeq \frac{1}{7.3}\,. \tag{12.9}$$

One could take into account that the neutron decays are increasing the number of protons too, but that's a small correction. It is quite a bizarre coincidence that the neutron half-life is so comparable to the time it takes the nuclei to form; if it had been much shorter all neutrons would decay and only hydrogen could form.

In the early Universe, the only elements produced in any significant abundance are hydrogen and helium-4. The latter is produced because it is the most stable light nucleus, and the former because there aren't enough neutrons around for all the protons to bind with and so some protons are left over. We can therefore get an estimate of their relative abundance, normally quoted as the fraction of the mass (not number density) of the Universe which is in helium-4. Since every helium nucleus contains 2 neutrons (and hydrogen contains none), all neutrons end up in helium and the number density of helium-4 is $N_{\mathrm{He-4}} = N_{\mathrm{n}}/2$. Each helium nucleus weighs about four proton masses, so the fraction of the total mass in helium-4, known as Y_4, is

$$Y_4 \equiv \frac{2N_{\mathrm{n}}}{N_{\mathrm{n}} + N_{\mathrm{p}}} = \frac{2}{1 + N_{\mathrm{p}}/N_{\mathrm{n}}} \simeq 0.24\,. \tag{12.10}$$

So this simple treatment tells us that about 24% of the matter in the Universe is in the form of helium-4. Note that this is the mass fraction; since helium-4 weighs four times as much as hydrogen, it means there is one helium-4 nucleus for every 14 hydrogen ones.

A more detailed treatment involves keeping track of a whole network of nuclear reactions, and carefully analyzing the balance between nuclear reaction rates and the expansion rate of the Universe. This typically gives answers in the range 23% to 24% helium-4, with almost all the rest in hydrogen. This reaction network also allows one to estimate the trace abundances of all the other nuclei which form in the early Universe. These are deuterium, helium-3 and lithium-7. By mass, these contribute about 10^{-4}, 10^{-5} and 10^{-10} respectively.

12.2 Comparing with observations

Remarkably, all of these element abundances can be measured, even that of lithium-7. This allows an extraordinarily powerful test of the Hot Big Bang model, encompassing ten orders of magnitude in abundance. There turn out to be only two important input parameters which affect the abundances.

1. The number of massless neutrino species in the Universe, which affects the expansion temperature–time relation and hence the way in which nuclear reactions go out of thermal equilibrium. So far we have assumed there are three neutrino types as in the Standard Model of particle interactions, but other numbers are possible in principle.

2. The density of baryonic matter in the Universe, from which the nuclei are composed. If the density of baryons were changed, it is reasonable to imagine that the details of how they form nuclei are changed. The absolute density of baryons, $\rho_{\rm B}$, is what matters. Normally this is expressed using the density parameter, and since the critical density $\rho_{\rm c}$ has a factor h^2 in it, that means that it is the combination $\Omega_{\rm B}h^2$ which is constrained.

An impressive success of the Big Bang model is that it was found that agreement with the observed element abundances could only be obtained if the number of massless neutrino species is three, which corresponds exactly to the three species (electron, muon and tau) we know to exist. When first obtained in the late 1980s, this result had no independent support, but since then the LEP experiment at CERN has confirmed the result of there being only three light neutrino species, based on the decay of the Z_0 particle. This is powerful indirect evidence that the predicted cosmic neutrino background does exist.

Once we fix the number of neutrinos at three, that leaves only $\Omega_{\rm B}h^2$ as an input parameter. Figure 12.1 shows the predicted abundances as a function of this parameter. The Hot Big Bang theory can successfully reproduce the observed abundances of all the light elements, provided $\Omega_{\rm B}h^2$ lies in a relatively narrow range. As it happens, the measured lithium-7 abundance lies near a minimum in the model prediction as a function of $\Omega_{\rm B}h^2$.

As agreement with the observations is only available for a limited range of $\Omega_{\rm B}h^2$, we have a very tight bound on the amount of baryonic matter there can be in the Universe, as was already discussed in Chapter 9. The strongest constraints arise from measurements of the deuterium abundance by absorption of quasar light as it passes through primordial gas clouds. Those give the darker vertical band in Figure 12.1, corresponding to

$$0.016 \leq \Omega_{\rm B}h^2 \leq 0.024\,, \tag{12.11}$$

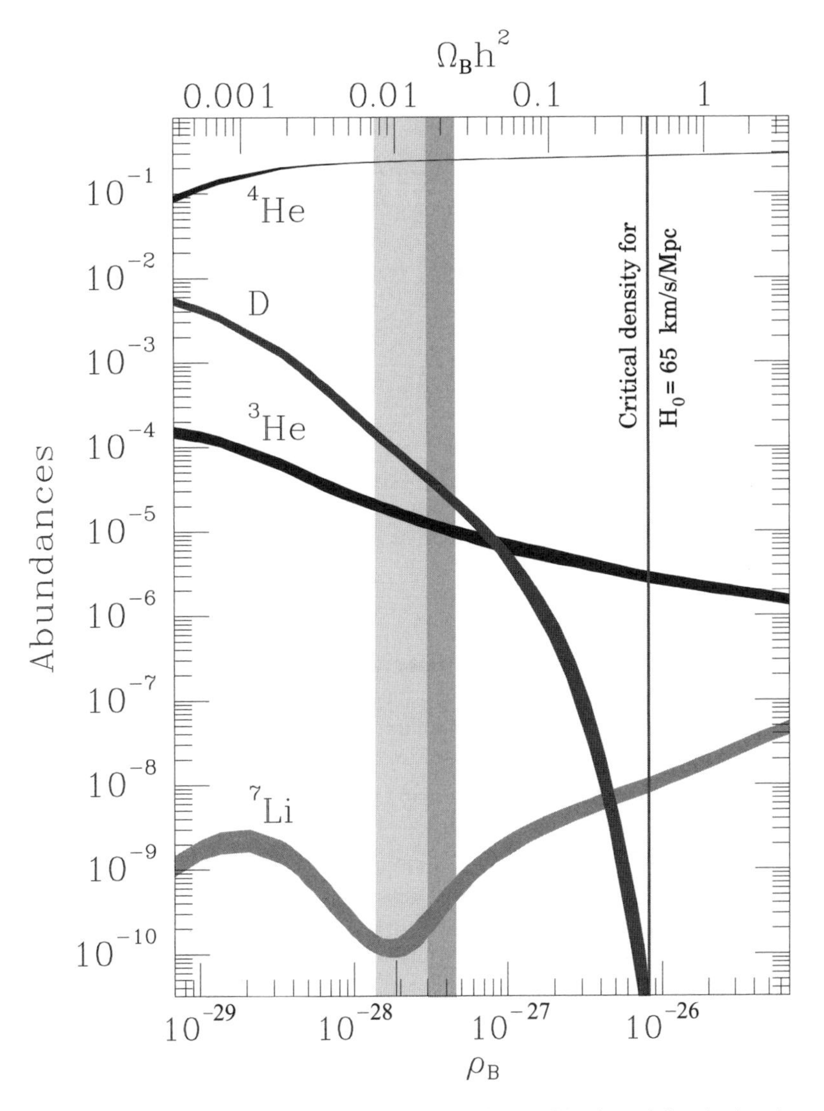

Figure 12.1 The predicted abundances of light nuclei, as a function of the absolute baryon density along the bottom or $\Omega_B h^2$ along the top. The width of the bands indicates the uncertainty in the predictions. The vertical bands show the range compatible with observations, while the vertical line shows the critical density. [From Schramm & Turner, Rev. Mod. Phys. **70**, 303 (1998), courtesy Michael Turner].

as already quoted in Chapter 9. This is a 95% confidence interval. Initially the deuterium measurements were quite controversial, though they have now become widely accepted; if they are ignored then the lower bound on Ω_B weakens quite a bit, as shown by the lighter vertical band (the allowed region being the combined bands in that case).

The figure also shows the critical density (assuming a particular value of h). This confirms, as discussed in Chapter 9, that it is impossible to have a critical density of bary-

Table 12.1 A comparison of nucleosynthesis and decoupling.

	Nucleosynthesis	Decoupling
Time	a few minutes	300 000 yrs
Temperature	10^{10} K	3000 K
Typical energy	1 MeV	1 eV
Process	Protons and neutrons form nuclei. Electrons remain free.	Nuclei and electrons form atoms.
Radiation	continues to interact with nuclei and electrons.	ceases interaction and forms microwave background.

onic matter unless the entire nucleosynthesis argument is somehow completely incorrect, despite giving such impressive answers. In fact, the upper limit from nucleosynthesis is some way below the observed density of matter in the Universe, giving strong support to the idea of non-baryonic dark matter.

Having measured the baryon density using nucleosynthesis, we then start to wonder about the density of anti-baryons. In particle physics, every particle has its anti-particle. However, it is believed that there are no significant quantities of anti-matter in our Universe; the annihilation signals of matter and anti-matter coming together are just too strong and would have already been seen. So our Universe possesses a matter–anti-matter asymmetry, quantified by the amount of baryonic matter within it. How this might arise is discussed in Advanced Topic 4.

12.3 Contrasting decoupling and nucleosynthesis

Because of the strong parallels between nucleosynthesis and decoupling, it is important to keep their properties distinct in your mind. The differences between the two arise in the huge difference in energy scales between atomic and nuclear processes, most disturbingly illustrated by the vastly different destructive powers of chemical and nuclear bombs.[1] The Universe is only hot enough to destroy nuclei for the first few minutes of its existence (until $t \simeq 400$ sec), while it remains capable of destroying atoms for more than a hundred thousand years. Table 12.1 summarizes the different scales and processes. In particular, remember that it is only decoupling which leads to the microwave background; after nucleosynthesis the photons are still able to interact with both nuclei and electrons.

[1]Don't be fooled by the incorrect terminology of 'atom bomb'.

Problems

12.1. By the standards of typical nuclear reactions, the neutron half-life of 614 seconds is extraordinarily long. What would be the consequence for light element production had this half-life instead been tiny (say a microsecond, for example)?

12.2. Imagine an alternate Universe where the neutron half-life is 100 seconds rather than 614 seconds. Estimate the fraction of the total mass of baryonic matter in the form of helium once nucleosynthesis is over in such a Universe.

12.3. Assuming that the Universe is charge neutral, how many electrons are there per baryon?

12.4. Which of decoupling and nucleosynthesis do you feel is the stronger test of the Hot Big Bang cosmology, and why?

Chapter 13

The Inflationary Universe

We now leave the well-established and understood topics in cosmology in order to discuss something more speculative. The idea in question is **cosmological inflation**, which was invented in 1981 and remains a hot research topic in modern cosmology. Inflation is not a replacement for the Hot Big Bang theory, but rather an extra add-on idea which is supposed to apply during some very early stage of the Universe's expansion. By the time the Universe has reached the ages we have already discussed, inflation is supposed to be long since over and the standard Big Bang evolution restored, in order to preserve the considerable successes we have already discussed, such as the microwave background and nucleosynthesis.

13.1 Problems with the Hot Big Bang

Before describing the idea of inflation, I will cover some of the historical motivations which led to its introduction. They arise from the realization that, despite all its successes, there remain some unsatisfactory aspects to the Hot Big Bang theory.

13.1.1 The flatness problem

The flatness problem is the easiest one to understand. We have learned that the Universe possesses a total density of material, $\Omega_{\rm tot} = \Omega_0 + \Omega_\Lambda$, which is close to the critical density. Very conservatively, it is known to lie in the range $0.5 \leq \Omega_{\rm tot} \leq 1.5$. In terms of geometry, that means that the Universe is quite close to possessing the flat (Euclidean) geometry.

We have seen that the Friedmann equation can be rewritten as an equation showing how $\Omega_{\rm tot}$ varies with time. Adding modulus signs to equation (7.4), this is

$$|\Omega_{\rm tot}(t) - 1| = \frac{|k|}{a^2 H^2} \,. \tag{13.1}$$

We know from this that if $\Omega_{\rm tot}$ is precisely equal to one, then it remains so for all time. But what if it is not?

Let's consider the situation where we have a conventional Universe (matter or radiation dominated) where the normal matter is more important than the curvature or cosmological constant term. Then we can use the solutions ignoring the curvature term, equations (5.15) and (5.19), to find

$$a^2H^2 \propto t^{-1} \qquad \text{radiation domination;} \tag{13.2}$$

$$a^2H^2 \propto t^{-2/3} \qquad \text{matter domination.} \tag{13.3}$$

So we have

$$|\Omega_{\rm tot} - 1| \propto t \qquad \text{radiation domination;} \tag{13.4}$$

$$|\Omega_{\rm tot} - 1| \propto t^{2/3} \qquad \text{matter domination.} \tag{13.5}$$

In either case, the difference between $\Omega_{\rm tot}$ and 1 is an *increasing* function of time. That means that the flat geometry is an *unstable* situation for the Universe; if there is any deviation from it then the Universe will very quickly become more and more curved. Consequently, for the Universe to be so close to flat even at its large present age means that at very early times it must have been extremely close to the flat geometry.

An alternative way to see this is to remember that the densities of matter and radiation reduce with expansion as $1/a^3$ and $1/a^4$ respectively. These are both faster reductions than the curvature term k/a^2. So if the curvature term is not to totally dominate in the present Universe, it must have begun much smaller than the other terms.

The equations for $|\Omega_{\rm tot} - 1|$ derived above stop being valid once the curvature or cosmological constant terms are no longer negligible, since we used the $a(t)$ solutions for the flat geometry to derive them. But they are fine to give us an approximate idea of what the problem is. For extra ease let's assume that the Universe always has only radiation in it. Using the equations above, we can ask how close to one the density parameter must have been at various early times, based on the constraint today ($t_0 \simeq 4 \times 10^{17}$ sec).

- At decoupling ($t \simeq 10^{13}$ sec), we need $|\Omega_{\rm tot} - 1| \leq 10^{-5}$.
- At matter–radiation equality ($t \simeq 10^{12}$ sec), we need $|\Omega_{\rm tot} - 1| \leq 10^{-6}$.
- At nucleosynthesis ($t \simeq 1$ sec), we need $|\Omega_{\rm tot} - 1| \leq 10^{-18}$.
- At the scale of electro-weak symmetry breaking, which corresponds to the earliest known physics ($t \simeq 10^{-12}$ sec), we need $|\Omega_{\rm tot} - 1| \leq 10^{-30}$.

Written out in long hand, that means we know that at nucleosynthesis, an era we are supposed to understand very well indeed, the density parameter must have lain within the range $0.999999999999999999 \leq \Omega_{\rm tot} \leq 1.000000000000000001$!! Out of all the possible values that it might have had, this seems a very restrictive range. Any other value would lead to a Universe extremely different to that which we see.

The easiest way out of this dilemma is to suppose that the Universe must have precisely the critical density. But on the face of it there seems no reason to prefer this choice over any other. What would be nice would be an explanation of such a value.

Regardless of whether or not we understand the physical origin of these numbers, they are an observed fact. One useful thing they tell us is that the Universe is very close to spatial flatness at decoupling and at nucleosynthesis, which means that it is always a good approximation to set $k = 0$ in the Friedmann equation when describing those phenomena.

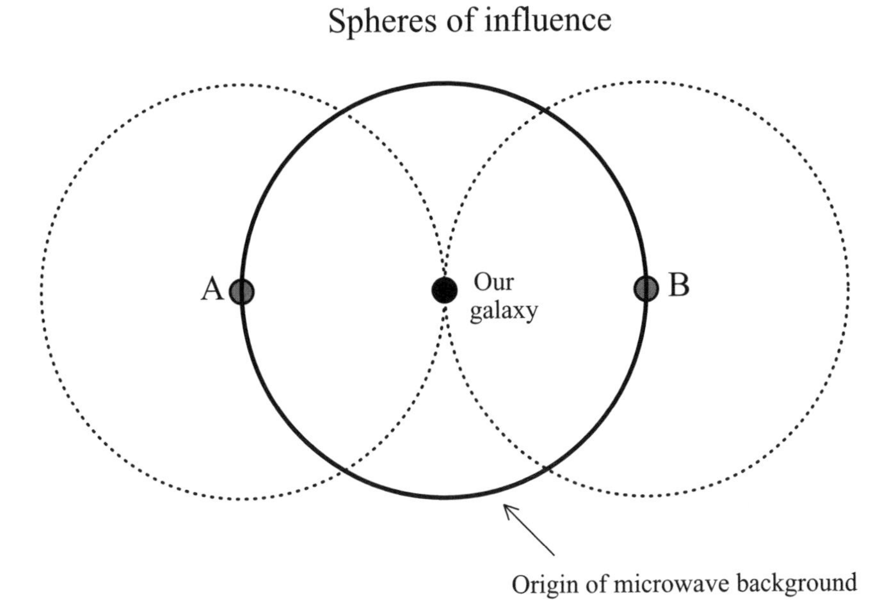

Figure 13.1 An illustration of the horizon problem. We receive microwave radiation from points **A** and **B** on opposite sides of the sky. These points are well separated and would not have been able to interact at all since the Big Bang — the dotted lines indicate the extent of regions able to influence points **A** and **B** by the present — far less manage to interact by the time the microwave radiation was released. So in the Hot Big Bang model it is impossible to explain why they have the same temperature to such accuracy.

13.1.2 The horizon problem

The horizon problem is the most important problem with the Hot Big Bang model, and refers to communication between different regions of the Universe. The crucial ingredient is that the Universe has only a finite age, and so even light can only have travelled a finite distance by any given time. As I have remarked, the distance which light could have travelled during the lifetime of the Universe gives rise to a region known as the observable Universe. This is the region we can actually see, and is always finite regardless of whether or not the Universe as a whole is finite or infinite.

One of the most important properties of the microwave background is that it is very nearly isotropic. That is, light seen from all parts of the sky possesses, to very great accuracy, the same temperature of 2.725 K. Being at the same temperature is the characteristic of thermal equilibrium, and so this observation is naturally explained if different regions of the sky have been able to interact and move towards thermal equilibrium. Unfortunately, the light we see from opposite sides of the sky has been travelling towards us since decoupling, close to the time of the Big Bang itself. Since the light has only just reached us, it can't possibly have made it all the way across to the opposite side of the sky. Therefore there has not been time for two regions on opposite sides of the sky to interact in any way, and so one cannot claim that the regions have the same temperature because they have interacted and established thermal equilibrium. This is illustrated in Figure 13.1.

In fact the problem is even worse, because the microwaves have been travelling uninterrupted since decoupling. Regions would have had to interact and thermalize even before then, by which time light could only have travelled a very short distance indeed — the observable Universe is much smaller at early times as light could have travelled much less far. So it transpires that even regions which appear quite close to each other on the sky (any points more than about a degree or two apart — see Problem 13.4) would not have been able to interact before decoupling to establish thermal equilibrium.

The final twist in the tail, which elevates this to a problem of extreme relevance, is that actually the microwave background is *not* perfectly isotropic, but instead exhibits small fluctuations (about one part in one hundred thousand) as detected by the COBE satellite. These irregularities are thought to represent the 'seeds' from which structure in the Universe grows, as described in Advanced Topic 5. For the same reason that one cannot thermalize separated regions, one also cannot create an irregularity. So in the standard Big Bang theory one cannot have a theory allowing the generation of the seed perturbations — they have to be there already.

13.1.3 Relic particle abundances

Another mystery arises from combining the Hot Big Bang model with modern ideas of particle physics. One of the curious things about the Universe is that it remained radiation dominated for so long, until an age of at least 1000 years. That is unexpected because the radiation density reduces with expansion as $1/a^4$, much faster than any other type of matter. If the Universe starts with just a very small amount of non-relativistic matter, then its slower reduction in density will rapidly bring it to prominence.

In fact, the particles in the Standard Model of particle interactions don't lead to any problems, because they interact strongly with radiation and thermalization stops them becoming too prominent. But modern particle physics throws up other particles. The most crucial in originally motivating inflation was a type of particle known as a magnetic monopole. Such particles are an inevitable consequence of models of unification of fundamental forces, the so-called Grand Unified Theories, and it is predicted that they were produced with a high abundance at a very early stage in the Universe. They are predicted to be extraordinarily massive; the Grand Unified Scale is thought to be around 10^{16} GeV, in comparison to the proton's puny 1 GeV or so. Such particles would be non-relativistic for almost all the Universe's history, giving them plenty of time to come to dominate over radiation. Since we know the Universe is not dominated by magnetic monopoles now, theories predicting them are incompatible with the standard Hot Big Bang model. This is further explored in Problem 13.5.

While magnetic monopoles were the relic particle thought most important at the time inflation was conceived, there are now several other kinds of relic particle also speculated to exist which would cause similar problems, going under such elaborate names as gravitinos and moduli fields.

13.2 Inflationary expansion

Alan Guth proposed **inflation** in 1981 as a solution to all of these problems. Stripped to its bare bones, inflation is defined as a period in the evolution of the Universe during which the scale factor was accelerating

$$\text{INFLATION} \quad \Longleftrightarrow \quad \ddot{a}(t) > 0\,. \tag{13.6}$$

Typically this corresponds to a very rapid expansion of the Universe.

Looking at the acceleration equation

$$\frac{\ddot{a}}{a} = -\frac{4\pi G}{3}\left(\rho + \frac{3p}{c^2}\right), \tag{13.7}$$

we see immediately that this implies $\rho c^2 + 3p < 0$. Since we always assume a positive density, this requires a negative pressure,

$$p < -\frac{\rho c^2}{3}\,. \tag{13.8}$$

Fortunately, modern particle physics ideas of symmetry breaking indicate ways in which this negative pressure can be brought about, described later in this chapter.

The classic example of inflationary expansion is a Universe possessing a cosmological constant Λ. This is equivalent to having a fluid with $p = -\rho c^2$ (see Section 7.2), which satisfies the condition above. We saw in Chapter 7 that the full Friedmann equation, including other matter terms and curvature, becomes

$$H^2 = \frac{8\pi G}{3}\rho - \frac{k}{a^2} + \frac{\Lambda}{3}\,. \tag{13.9}$$

If all the terms on the right-hand side are significant, then this is quite complex. Fortunately, though, the situation quickly becomes more simple, because the first two terms are rapidly reduced by the expansion while the last one remains constant. So after a while, only the cosmological constant term will be significant and we will have

$$H^2 = \frac{\Lambda}{3}\,. \tag{13.10}$$

Recalling that $H = \dot{a}/a$, this means

$$\dot{a} = \sqrt{\frac{\Lambda}{3}}\,a\,, \tag{13.11}$$

which, since Λ is a constant, has the solution

$$a(t) = \exp\left(\sqrt{\frac{\Lambda}{3}}\,t\right). \tag{13.12}$$

So when the Universe is dominated by a cosmological constant, the expansion rate of the Universe is much more dramatic than those we have seen so far.

After some amount of time, inflation must come to an end, with the energy in the cosmological constant being converted into conventional matter. One should think of this as a decay of the particles acting as the cosmological constant into normal particles. The Big Bang can then proceed just as before. Provided all this happens when the Universe was extremely young, none of the successes of the Hot Big Bang model are lost. In typical models the Universe is extremely young indeed when inflation is supposed to occur, perhaps around 10^{-34} sec which is about the time appropriate to the Grand Unification scale of 10^{16} GeV — see equation (11.12).

13.3 Solving the Big Bang problems

13.3.1 The flatness problem

Recall that we rewrote the Friedmann equation as

$$|\Omega_{\rm tot}(t) - 1| = \frac{|k|}{a^2 H^2} \,. \tag{13.13}$$

In the Big Bang theory, the problem was that this always increases with time, forcing $\Omega_{\rm tot}$ away from one.

Inflation reverses this state of affairs, because

$$\ddot{a} > 0 \quad \Longrightarrow \quad \frac{\mathrm{d}}{\mathrm{d}t}(\dot{a}) > 0 \quad \Longrightarrow \quad \frac{\mathrm{d}}{\mathrm{d}t}(aH) > 0 \,. \tag{13.14}$$

So the condition for inflation is precisely that which drives $\Omega_{\rm tot}$ towards one rather than away from one. In the special case of perfect exponential expansion, the approach is particularly dramatic

$$|\Omega_{\rm tot}(t) - 1| \propto \exp\left(-\sqrt{\frac{4\Lambda}{3}}\, t\right) . \tag{13.15}$$

The aim is to use inflation not just to force $\Omega_{\rm tot}$ close to one, but in fact to make it so extraordinarily close to one that even all the subsequent expansion between the end of inflation and the present is insufficient to move it away again, as shown in Figure 13.2. In the next section we'll see how much inflation that entails.

The standard analogy for this solution to the flatness problem is to imagine a balloon being very rapidly blown up, say to the size of the Sun; its surface would then look flat to us. The crucial difference inflation introduces compared to the usual Big Bang case is that the size of the portion of the Universe you can observe, given roughly by the Hubble length cH^{-1} (since H^{-1} is roughly the age of the Universe and c the maximum speed), does not change while this happens. So very quickly you are unable to notice the curvature of the surface. By contrast, in the Big Bang scenario the distance you can see increases more quickly than the balloon expands, so you can see more of the curvature as time goes by.

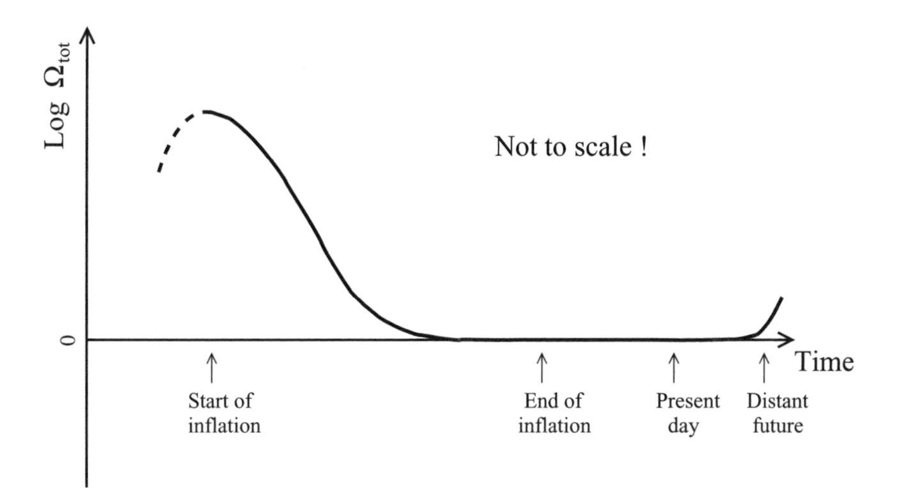

Figure 13.2 Possible evolution of the density parameter Ω_{tot}. There might or might not be a period before inflation, indicated by the dashed line. Inflation then drives $\log \Omega_{tot}$ towards zero (i.e. Ω_{tot} towards 1), either from above or below. By the time inflation ends Ω_{tot} is so close to one that all the evolution after inflation up to the present day is not enough to pull it away again. Only some time in the very distant future would it start to move away from one again.

Inflation predicts a Universe extremely close to spatial flatness. If one allows the possibility of a cosmological constant in the present Universe, then a flat Universe requires

$$\Omega_0 + \Omega_\Lambda = 1\,. \tag{13.16}$$

Current observations, particularly of cosmic microwave background anisotropies, strongly suggest that this condition is indeed satisfied. So far, then, this simple prediction of inflation stands up well to confrontation with observations.

13.3.2 The horizon problem

Inflation greatly increases the size of a region of the Universe, while keeping its characteristic scale, the Hubble scale, fixed. This means that a small patch of the Universe, small enough to achieve thermalization before inflation, can expand to be much larger than the size of our presently observable Universe, as shown in Figure 13.3. Then the microwaves coming from opposite sides of the sky really are at the same temperature because they were once in equilibrium. Equally, this provides the opportunity to generate irregularities in the Universe which can lead to structure formation.

Another way of expressing the resolution of the horizon problem is to say that, because of inflation, light can travel a much greater distance between the Big Bang and the time of decoupling than it can between decoupling and the present, reversing the usual state of affairs.

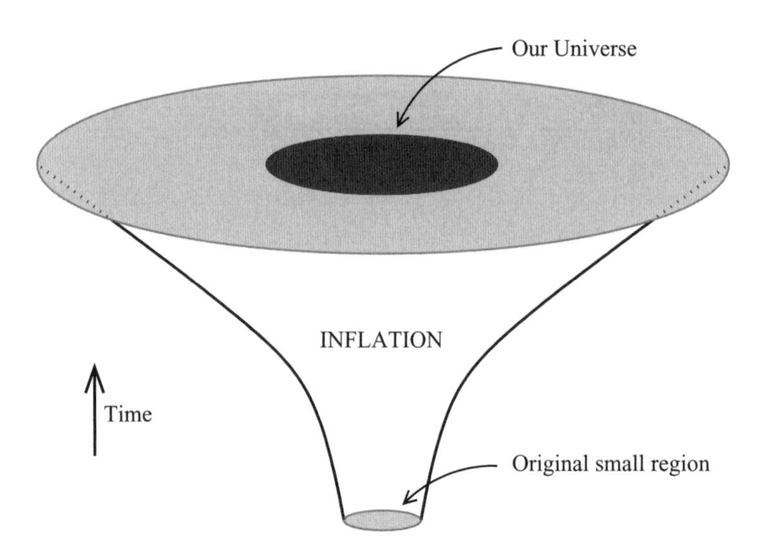

Figure 13.3 A schematic illustration of the inflationary solution to the horizon problem, with a small initial thermalized region blown up to encompass our entire observable Universe.

13.3.3 Relic particle abundances

The dramatic expansion of the inflationary era dilutes away any unfortunate relic particles, because their density is reduced by the expansion more quickly than the cosmological constant. Provided enough expansion occurs, this dilution can easily make sure that the particles are not observable today; in fact, rather less expansion is needed than to solve the other problems.

One important proviso though is that the decay of the cosmological constant which ends inflation must not regenerate the troublesome particles again. This means that the temperature which the Universe is at after inflation must not be too high, in order to make sure there is no new thermal production.

13.4 How much inflation?

We can use the flatness problem to estimate how much expansion is needed from inflation. I'll make the following simplifying assumptions, all of which could be relaxed for a better calculation.

- Inflation ends at 10^{-34} sec.
- The inflationary expansion is perfectly exponential.
- The Universe is perfectly radiation dominated all the way from the end of inflation to the present.
- The value of $\Omega_{\rm tot}$ near the start of inflation is not hugely different from one.
- For the sake of argument, assume the present value of $|\Omega_{\rm tot} - 1| \leq 0.1$.

The present age of the Universe is about 4×10^{17} sec. During radiation domination

$$|\Omega_{\rm tot}(t) - 1| \propto t\,, \tag{13.17}$$

so

$$|\Omega_{\rm tot}(t_0) - 1| \leq 0.1 \quad \Rightarrow \quad \left|\Omega_{\rm tot}(10^{-34}\,{\rm sec}) - 1\right| \leq 3 \times 10^{-53}\,. \tag{13.18}$$

During inflation H is constant, so

$$|\Omega_{\rm tot}(t) - 1| \propto \frac{1}{a^2}\,. \tag{13.19}$$

So the required value at the end of inflation can be achieved provided that during inflation a is increased by a factor of at least 10^{27}!! Incredibly, by the standards of what comes out of inflation model building this isn't much at all. Expansion by factors like 10^{10^8} are not uncommon!

This can all happen very quickly. Suppose for example that the characteristic expansion time, H^{-1}, is 10^{-36} sec. Then between 10^{-36} sec and 10^{-34} sec, the Universe would have expanded by a factor

$$\frac{a_{\rm final}}{a_{\rm initial}} \simeq \exp\left[H(t_{\rm final} - t_{\rm initial})\right] = e^{99} \simeq 10^{43}\,. \tag{13.20}$$

The exponential expansion is so dramatic that very large expansion factors drop out almost automatically.

13.5 Inflation and particle physics

The way I've discussed inflation, defining it as a period of accelerated expansion and showing that, for example, a cosmological constant can give such behaviour, is fine for developing an understanding of what inflation is and why it can solve the various cosmological problems. However, simply postulating a cosmological constant and claiming that it is able to decay away after having done its work is clearly a very *ad hoc* approach. A true model of inflation should contain a reasonable hypothesis for the origin of the cosmological constant, and a natural way of bringing inflation to an end.

To find such a model, we have to search the realms of particle physics. Remember that we must not spoil nucleosynthesis, so the very latest that inflation could have happened was when the Universe was one second old. We saw in Chapter 11 that this already corresponds to temperatures of over 10^{10} K, and in fact typical inflation models happen at much earlier times, and hence hotter environments, than that. In order to describe such extreme physical conditions, in which violent particle collisions are the norm, fundamental particle physics is required, and in particular theories of the fundamental interactions. Inflation is assumed to be driven by a new, as-yet-undiscovered, form of matter required by such theories.

A key idea is that of phase transitions. A phase transition corresponds to a dramatic change in the properties of a physical system as it is heated or cooled. Familiar examples are the freezing of water into ice, the lining up of domains in a cooled ferromagnet, or the onset of superconductivity or superfluidity at low temperatures. It is believed that the Universe itself will have undergone a series of phase transitions as it cooled, an example being when quarks first condensed to form hadrons.

A phase transition is a particularly dramatic event in the history of the Universe, a time when its properties change substantially. Phase transitions are controlled by an unusual form of matter known as a **scalar field**. Depending on the precise nature of the transition, scalar fields can behave with a negative pressure, and can satisfy the inflationary condition $\rho c^2 + 3p < 0$. That is, they behave like an effective cosmological constant. Once the phase transition comes to an end, the scalar field decays away and the inflationary expansion terminates, hopefully having achieved the necessary expansion by a factor of 10^{27} or more.

Inflation is currently a very active research field, and most of the study is carried out under the general hypothesis that inflation is driven by a scalar field. The hope is that eventually some specific particle physics phase transition can be identified which is likely to be the one giving inflation. Early work focussed on the Grand Unification phase transition, where the strong nuclear force first obtains an identity distinct from the electro-weak force (which will itself later split into the weak nuclear force and the electromagnetic interaction). This is believed to have happened at the very high energy of 10^{16} GeV, when the Universe was only 10^{-34} sec old, and was the example I used in working out the amount of inflation required.

More recently, attention has focussed on a different idea known as supersymmetry, already invoked in Chapter 9 to give a dark matter candidate. Supersymmetry postulates that every fundamental particle we know about, such as photons, electrons and quarks, has a partner particle with similar properties but with a higher mass. This higher mass makes them very difficult to create using particle accelerators, which is why they have yet to be seen in experiments (apart from the obvious possibility that they haven't been seen because they are a figment of particle physicists' imaginations). In the early Universe, the particles and their partners would have had very similar properties, and then a phase transition would lead to their present, more separate, identities. Currently, supersymmetric theories of particle physics appear the best prospect for creating models for the inflationary expansion. However, there are now a very large number of different models of inflation, and one of the goals of cosmology is to narrow this down to a favoured model or, alternatively, to disprove the inflation theory.

Problems

13.1. During standard Big Bang evolution, we have seen that $\Omega_{\rm tot}$ moves away from one unless its initial value was precisely one. Can $\Omega_{\rm tot}$ become infinite, and if so what does this mean?

13.2. Certain models of the early Universe permit an expansion rate $a \propto t^m$ where m is an arbitrary positive constant. What range of values of m corresponds to an inflationary expansion?

13.3. In a radiation-dominated Universe, the temperature and time are related by equation (11.12)

$$\left(\frac{1\,\mathrm{sec}}{t}\right)^{1/2} \simeq \frac{T}{2\times 10^{10}\,\mathrm{K}}\,.$$

At what time was the temperature $3 \times 10^{25}\,\mathrm{K}$?

Suppose that at that time Ω_{tot} were a bit less than one, so that the Universe quickly became curvature dominated, with expansion law $a(t) \propto t$. How would the above equation be changed, and how old would the Universe have been when its temperature fell to 3K?

13.4. In this question, assume throughout that the Universe is matter dominated with critical density, so that $a(t) \propto t^{2/3}$. Take the present Hubble constant to be $100\,\mathrm{km\,s^{-1}\,Mpc^{-1}} \simeq 10^{-10}\,\mathrm{yr}^{-1}$. As we have seen, the age of the Universe is given fairly accurately by the Hubble time, H^{-1}. Estimate how far light can have travelled since the Big Bang, given that the speed of light is $c \simeq 3 \times 10^{-7}\,\mathrm{Mpc\,yr}^{-1}$.

The microwave background radiation has been travelling towards us uninterrupted since decoupling, when the Universe was one thousandth its present size. Compute the value of the Hubble parameter at the time of decoupling. How far could light have travelled in the time up to decoupling?

Between decoupling and the present, the distance that the light travelled up to the time of decoupling has been stretched by the subsequent expansion. What would be its physical size today?

Assuming that the distance to the origin of the microwave background, known as the last-scattering surface, is given by your answer to the first part of this question, what angle is subtended by the distance light could have travelled before decoupling? What is the physical significance of this angle?

13.5. Magnetic monopoles behave as non-relativistic matter. Suppose that at a temperature corresponding to the Grand Unified era, about $3\times10^{28}\,\mathrm{K}$, magnetic monopoles were created with a density of $\Omega_{\mathrm{mon}} = 10^{-10}$. Assuming that the Universe has a critical density and is radiation dominated, what was the temperature when the density of monopoles equalled that of the radiation?

In the present Universe, $T \simeq 3\,\mathrm{K}$. Compute the value $\Omega_{\mathrm{mon}}/\Omega_{\mathrm{rad}}$ would have at the present day. Is this ratio compatible with observations?

13.6. Consider the situation of Problem 13.5. If we have a period of inflation, the monopole density still reduces as $\rho_{\mathrm{mon}} \propto 1/a^3$, but the total density, dominated by the cosmological constant, remains fixed. Since that density will be converted to radiation after inflation, we can imagine that the radiation density remains constant during inflation. How much inflationary expansion is necessary so that the present density of monopoles matches that of radiation?

Chapter 14

The Initial Singularity

Having completed our diversion into structures in the Universe, we can now return to our tracing of the Universe's history. We began by describing its present state. We then studied the successes of decoupling and nucleosynthesis, and then more speculatively considered inflation as a theory of the possible early evolution. We are now led to the ultimate question: if the Universe has always been expanding, must it have had a beginning?

Let's begin with a historical perspective. In the 1960s it was believed that any conceivable form of matter would obey a condition known as the **strong energy condition**,

$$\rho c^2 + 3p \geq 0 \,. \tag{14.1}$$

Under this assumption, we see immediately from the acceleration equation

$$\frac{\ddot{a}}{a} = -\frac{4\pi G}{3}\left(\rho + \frac{3p}{c^2}\right), \tag{14.2}$$

that the Universe was always decelerating. This allows us to prove, as follows, that if the Universe is homogeneous then it must have had a beginning.

Let's first assume that the Universe experiences no deceleration as it expands. Then, since we know it is expanding now, we get

$$\dot{a} = \text{const} \quad \Longrightarrow \quad a(t) = \text{const} \times (t - t_{\text{min}})\,, \tag{14.3}$$

which implies that $a(t)$ becomes zero at $t = t_{\text{min}}$. We can evaluate the integration constant t_{min} from the present expansion rate $H = \dot{a}/a$, getting

$$t_0 - t_{\text{min}} = H_0^{-1} = 9.77h^{-1} \times 10^9 \, \text{yrs} \,. \tag{14.4}$$

So, with no deceleration, the Universe would have an age equal to the Hubble time H_0^{-1}, as we found in Chapter 8. This is shown in Figure 14.1 as the dotted line.

Now, in reality the strong energy condition guarantees that the Universe is decelerating. That means that the true $a(t)$ must be curving downwards, while having the same slope at the present time t_0, again as shown in Figure 14.1. The dotted line is the tangent to

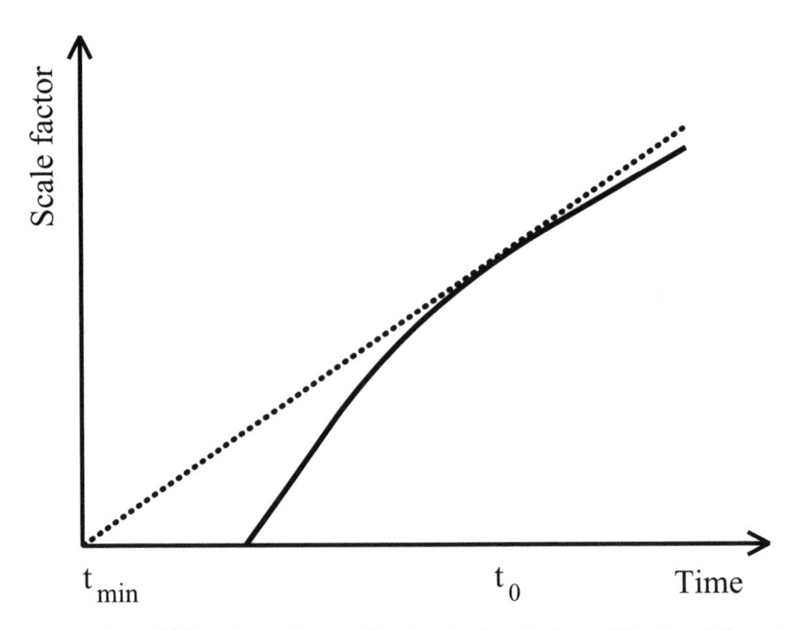

Figure 14.1 The solid line shows the true (decelerating) scale factor. The dotted line extrapolated back from the present shows the earliest possible time that the scale factor can have been zero.

the solid line at t_0. As you see, because the true $a(t)$ is curving down, it must intersect with the x-axis, $a = 0$, at some time *later* than $t_{\rm min}$. So, if the strong energy condition is obeyed, a homogeneous Universe must have had a zero scale factor at some time in the past more recent than H_0^{-1} before the present. This time is known as the **Big Bang**. At the time of the Big Bang, all the material in the Universe is crushed into a point of infinite density, and physical laws as we know them break down. For that reason, the Big Bang is also known as the **initial singularity**.

This argument may not apply directly to our Universe, however, as the presence of a cosmological constant breaks the strong energy condition. In Chapter 7 it was noted that if Λ is large enough, there are models with no Big Bang. However those bounce models are ruled out by observations of high-redshift objects, and so the existence of Λ in the present Universe is not thought to evade the above argument.

However, in the early nineteen sixties it was believed that this didn't necessarily imply an initial singularity to the Universe, because the argument depends on the assumption that the Universe is homogeneous and isotropic, which we know not to be absolutely true. It was believed that if the Universe was not perfectly isotropic, then during the collapse (thinking backwards in time again!) the irregularities might grow, and that the Universe might somehow 'miss' the crunch point and expand out again. It was also briefly believed that pressure might save the day, though as we have seen pressure actually increases the gravitational force. It therefore came as a big shock when Penrose and Hawking managed to prove several *singularity theorems*, which demonstrated that the existence of the initial singularity is extremely generic, under the assumption of the strong energy condition. Their results implied that there was indeed a 'Big Bang'.

Since then however the situation has become a lot less clear, because it has become widely accepted that a crucial plank of the proof, the strong energy condition, need not always be obeyed. Indeed, we have seen that inflation relies on its violation. That said, there might still be a more general, as-yet-undiscovered, generalization of the singularity theorems which doesn't require the assumption of the strong energy condition. Considerable work recently has gone into the investigation of the possibility of non-singular cosmologies, with only limited success. It is easy to find solutions which don't have singularities [for example, the perfect inflationary solution $a(t) \propto \exp(Ht)$ is perfectly finite back to $t = -\infty$]; the principal problem is that these situations don't seem to be stable solutions.

Finally, all the above discussion is based on classical physics. However, it is widely believed that when the Universe exceeded a certain density, quantum effects must have been important even for gravitational physics. This should happen when the energy scale is at or above the **Planck scale**, which is a characteristic scale formed from the three fundamental constants G, $\hbar$ and c, in such a way as to obtain the dimensions of energy

$$E_{\rm Pl} \equiv \sqrt{\frac{\hbar c^5}{G}} = 1.22 \times 10^{19}\,\mathrm{GeV}\,. \tag{14.5}$$

This density would have been achieved when the Universe's age equalled the Planck time

$$t_{\rm Pl} \equiv \sqrt{\frac{\hbar G}{c^5}} = 5.39 \times 10^{-44}\,\mathrm{sec}\,. \tag{14.6}$$

This is the earliest time we have seen mentioned in the entire book, earlier even than the time at which inflation is thought to take place. The mechanism for driving inflation is therefore part of classical, not quantum, physics.

The merger of gravitational physics with quantum physics is known as **quantum gravity**, but unfortunately at present we have no firm picture of what a quantum gravity theory might entail. In particular, Einstein's general relativity appears inconsistent with quantum mechanics. One candidate for a reconciliation is superstring theory (and its more modern variant M-theory). The implications of these ideas for the nature of the Big Bang itself have yet to be understood.

The question of whether our Universe really did experience a 'Big Bang' therefore remains open, though we do know that if anything strange happened to prevent it, it must be a property of the very earliest stages of the Universe's evolution. One possibility is that the Universe originated by quantum tunnelling, somewhat in the manner of radioactive decay freeing an alpha particle from a nucleus. The most puzzling aspect of that is that since time and space do not exist independently of the Universe, the tunnelling must be from nothing. Meaning not just empty space, but from a state where space and time had yet to exist!

Problem

14.1. In a radiation-dominated Universe, what would be the temperature at the Planck time?

Chapter 15

Overview: The Standard Cosmological Model

This book has provided an introductory account of many of the most important topics in modern cosmology. As you've seen, there remain a number of unresolved issues, many of which should however be accessible to observation over the next decade, as astronomical instrumentation continues to improve.

Here is a brief summary of what we've learned. In combination, it adds up to a Standard Cosmological Model, with an almost universal consensus amongst cosmologists that it represents our best understanding of the observational data, the first time in the history of cosmology that such a consensus has existed. Where specific observational results are quoted, remember that they were written in late 2007, and may have changed. Check out

`http://astronomy.susx.ac.uk/~andrewl/cosbook.html`

where, if my computer account is still valid, you can find some more up-to-date opinions. In creating this second edition I was struck by how much rewriting this chapter required from the 1998 first edition, so if you are reading this long after 2007 it should be well worth a look.

Expansion

The expansion of the Universe is a long-established fact. However, the difficulty of measuring distances to galaxies, independent of their redshift, has meant that fixing the present rate of expansion, the Hubble constant, has been a lengthy and exhausting process stretching back over fifty years. Nevertheless, through efforts including the Hubble Space Telescope Key Project, and more generally through combination of the large array of evidence supporting the standard cosmological model, it now seems certain that $h \simeq 0.7$ within ten percent accuracy at 95% confidence.

An interesting question is whether or not the expansion is accelerating or decelerating, as measured by q_0 (Chapter 6). Acceleration is the conclusion from studies of distant type Ia supernovae, and carries independent support from observations of structure formation. It seems that our Universe is presently experiencing an inflationary expansion!

Geometry

There has long been theoretical support for a flat geometry from the inflationary cosmology, which generates flatness while resolving other cosmological issues. Only recently has this become strongly supported by observations, in particular of the angular size of features in the cosmic microwave background. While it is impossible to ever prove that the Universe is precisely flat, all indications are that it is at least extremely close, with the sum of all density parameters (including the cosmological constant) adding to one to within a few percent accuracy.

Age

Results from the Hipparcos satellite have brought what were formerly alarmingly high age estimates for globular clusters down to much more comfortable values. At the same time, the introduction of the cosmological constant has pushed the predicted age of the Universe significantly upwards. The Standard Cosmological Model predicts a Universe which is about fourteen billion years old, and the long-standing paradox that the oldest stars seemed older than the Universe appears to have vanished.

Fate

In the currently-favoured cosmological models, the Universe survives forever rather than recollapsing, and indeed the inferred cosmological constant is leading to an accelerated expansion at present. Nevertheless, it is dangerous to try and second-guess what physics might take over in future; the cosmological constant may be a transient phenomenon, as was the similar quantity believed to have driven inflation in the early Universe, and so the acceleration may one day cease. If a small negative cosmological constant were ever to appear, it could promote recollapse in the future.

Contents

The Universe contains several different kinds of material.

Radiation: The amount of radiation is accurately given from the cosmic microwave background temperature as $\Omega_{\mathrm{rad}} h^2 \simeq 2.47 \times 10^{-5}$.

Relativistic: The cosmic neutrino background cannot be observed directly, but plausible assumptions give the minimum possible combined energy density in photons and neutrinos as $\Omega_{\mathrm{rel}} h^2 \simeq 4 \times 10^{-5}$, if neutrino masses are all negligible. More likely is that neutrinos do have significant mass, and if so present cosmological limits permit a neutrino density up to around $\Omega_\nu = 0.01$.

Baryons: Nucleosynthesis indicates that baryons make up about 4 percent of the critical density, with a dependence on the value of h, in order to match the observed abundances of light elements. Cosmic microwave background anisotropy studies given a similar value, though presently with larger uncertainty.

Dark matter: There are so many separate pieces of supporting evidence that the case for the existence of dark matter is overwhelming. However, the actual density of dark matter is quite uncertain, though believed to be around $\Omega_{\rm dm} \simeq 0.25$, and the actual constitution totally unknown. Cold dark matter is preferred to hot, though a subdominant component of hot dark matter could still be present.

Cosmological constant: Current observations indicate that the energy density of the cosmological constant dominates the Universe, driving accelerated expansion. However, it is unable to replace all the dark matter since by definition its density is constant everywhere, whereas dark matter needs to be concentrated into galaxy halos to explain the rotation curves. There is some speculation that the cosmological constant might be a transient phenomenon and/or exhibit slow variation, in which case the effective cosmological 'constant' is often described as **quintessence**. Various models for quintessence exist, none being compelling.

Early history

Inflation is a compelling and observationally viable idea, but like all early Universe topics must await improved observations before it can be seriously tested.

Outlook

As I advertized at the start of the book, the Hot Big Bang cosmology is an impressive framework within which we are able to interpret the many kinds of observations we are now able to make. Five pieces of strong evidence in its favour stand out — the expansion of the Universe, the predicted age of the Universe, the existence and thermal form of the cosmic microwave background, the relative abundances of the light elements predicted by cosmic nucleosynthesis, and the ability to predict the observed structures in the galaxy distribution and cosmic microwave background. Cosmologists are in the process of establishing and verifying the Standard Cosmological Model as described above. Barring surprises, we are in the process entering the era of precision cosmology, where the basic parameters describing our Universe will be determined to the best possible accuracy, hopefully the percent level in many cases. This is the principal goal of cosmology over the next decade.

Advanced Topic 1

General Relativistic Cosmology

Prerequisites: Chapters 1 to 4

For those readers who have experienced some general relativity, this chapter outlines the construction of cosmological models using relativity. As far as we are presently aware, general relativity gives an excellent description of gravitational physics and is normally considered the correct setting for discussing cosmological models.

An important idea is the **metric** of space-time, which describes the physical distance between points, and the metric is important both for correctly interpretting the geometry of the Universe and to fully understand ideas of luminosities and distances in cosmology. Those are discussed further in Advanced Topic 2, and play an important role in evidence for the cosmological constant as discussed in Chapter 7. This chapter will also outline the derivation of the crucial Friedmann and fluid equations using general relativity, giving rise to the same equations as derived using Newtonian techniques in Chapter 3.

A1.1 The metric of space-time

In general relativity, the fundamental quantity is the metric which describes the geometry of space-time, by giving the distance between neighbouring points in space-time. To build intuition, we consider first the metric of a flat piece of paper, upon which points can be specified by coordinates x_1 and x_2. The distance ds between two points is given by Pythagoras's Theorem

$$\Delta s^2 = \Delta x_1^2 + \Delta x_2^2 \,, \tag{A1.1}$$

where Δx_1 and Δx_2 are the separations in the x_1 and x_2 coordinates. Now suppose we replace the paper by a rubber sheet and let it expand. If our coordinate grid x_1–x_2 expands with the sheet, then the physical distance between points grows with time, and if the expansion is uniform (i.e. independent of position) we can write

$$\Delta s^2 = a^2(t)\left[\Delta x_1^2 + \Delta x_2^2\right] \,, \tag{A1.2}$$

where $a(t)$ measures the rate of expansion. The coordinates x_1 and x_2 are comoving coordinates, exactly as described in Chapter 3.

In the simple example above, Δs indicated only the spatial distance between points. However, in general relativity we are interested in the distance between points in four-dimensional space-time, and we must also allow for the possibility that space-time might be curved. The separation can be written as

$$ds^2 = \sum_{\mu,\nu} g_{\mu\nu} dx^\mu dx^\nu \,, \tag{A1.3}$$

where $g_{\mu\nu}$ is the metric, μ and ν are indices taking the values 0, 1, 2 and 3, x^0 is the time coordinate and x^1, x^2 and x^3 are the three spatial coordinates. In general the metric is a function of the coordinates (indeed, to describe a curved space-time there must be some such dependence), and the distances are written in infinitesimal notation as once space-time is curved it only makes sense to give the distance to nearby points.

Fortunately, this complicated situation can immediately be dramatically simplified by imposing the cosmological principle that, at a given time, the Universe should not have any preferred locations. This requires that the spatial part of the metric has a constant curvature, a condition satisfied for example by a flat metric which has zero curvature everywhere. However this is not the most general possibility; the most general spatial metric which has constant curvature can be shown to be

$$ds_3^2 = \frac{dr^2}{1-kr^2} + r^2\left(d\theta^2 + \sin^2\theta\, d\phi^2\right) \,, \tag{A1.4}$$

where ds_3 refers only to the spatial dimensions, and spherical polar coordinates have been used. Here k is an undetermined constant which measures the curvature of space. The possibilities k positive, zero or negative correspond to the three possible spatial geometries spherical, flat or hyperbolic respectively, as described in Chapter 4.

Having found the most general spatial metric, we now need to incorporate it into a space-time. The only further dependencies we can put in are time dependences; in particular we can allow the space to grow or shrink with time. This leads us to the **Robertson–Walker metric**

$$ds^2 = -c^2dt^2 + a^2(t)\left[\frac{dr^2}{1-kr^2} + r^2\left(d\theta^2 + \sin^2\theta\, d\phi^2\right)\right] \,, \tag{A1.5}$$

where $a(t)$ is the **scale factor of the Universe**. It looks as though there could also have been a function of time $b^2(t)$ before the dt^2, but it could be removed by redefining the time coordinate as $dt \to dt' = b(t)dt$; general relativity tells us we are allowed to use any coordinate system and so this extension would be no more general than the form given.

A1.2 The Einstein equations

The metric evolves according to Einstein's equation

$$R^\mu_{\ \nu} - \frac{1}{2} g^\mu_{\ \nu} R = \frac{8\pi G}{c^4} T^\mu_{\ \nu} \,, \tag{A1.6}$$

where T^{μ}_{ν} is the energy–momentum tensor of any matter which is present, and R^{μ}_{ν} and R are the Ricci tensor and scalar respectively, which give the curvature of space-time. As the energy–momentum tensor is assumed symmetric there are potentially ten Einstein equations (the number of independent components of a 4×4 symmetric matrix). If, as here, the metric has additional symmetries, the number of independent Einstein equations may be much less.

Einstein's equation tells us how the presence of matter curves space-time, and so we need to describe the matter under consideration. The possible constituents of the Universe considered in this book are all examples of so-called perfect fluids, meaning that they have no viscosity or heat flow. Perfect fluids have energy–momentum tensor

$$T^{\mu}_{\nu} = \mathrm{diag}\left(-\rho c^2, p, p, p\right), \tag{A1.7}$$

where ρ is the mass density and p the pressure.

For this metric, there are two independent Einstein equations, the time–time one and the space–space one. The derivation is too lengthy to reproduce here, but can be found in any good general relativity textbook. The time–time Einstein equation gives precisely the Friedmann equation

$$\left(\frac{\dot{a}}{a}\right)^2 + \frac{kc^2}{a^2} = \frac{8\pi G}{3}\rho, \tag{A1.8}$$

exactly as equation (3.10) except for the interpretation of k. The second Einstein equation doesn't quite look like those we've seen so far, being

$$2\frac{\ddot{a}}{a} + \left(\frac{\dot{a}}{a}\right)^2 + \frac{kc^2}{a^2} = -8\pi G\frac{p}{c^2}, \tag{A1.9}$$

but if we subtract the Friedmann equation from it we get precisely the acceleration equation, equation (3.18). We can then derive the fluid equation from these two, just by reversing the way we obtained the acceleration equation from the Friedmann and fluid equations. In this way, we obtain our equations from general relativity. The equations we are using are exact; for homogeneous Universes Newtonian gravity, normally only an approximation to general relativity, happens to yield exactly the right result.

There is a slightly more direct route to the fluid equation, taking advantage of the fact that general relativity automatically encodes energy conservation. This can be written

$$T^{\mu}_{\nu;\mu} = 0, \tag{A1.10}$$

where the semicolon is a covariant derivative and a summation over the repeated μ index is assumed (the Einstein summation convention). Although this is really four equations (ν being any of the four space–time coordinates), only the time component gives a non-trivial equation. Writing out the covariant derivative using the Christoffel symbols $\Gamma^{\alpha}_{\ \beta\gamma}$ gives

$$T^{\mu}_{\nu,\mu} + \Gamma^{\mu}_{\ \alpha\mu} T^{\alpha}_{\nu} - \Gamma^{\alpha}_{\ \nu\mu} T^{\mu}_{\alpha} = 0, \tag{A1.11}$$

where a comma is an ordinary derivative. For the $\nu = 0$ component, remembering that T^μ_ν is diagonal, the relevant Christoffel symbols are

$$\Gamma^0{}_{00} = 0 \quad ; \quad \Gamma^1{}_{01} = \Gamma^2{}_{02} = \Gamma^3{}_{03} = \frac{\dot{a}}{a} \,. \tag{A1.12}$$

Substituting them in, keeping careful track of the summation over repeated indices, gives

$$\dot{\rho} + 3\frac{\dot{a}}{a}\left(\rho + \frac{p}{c^2}\right) = 0 \,. \tag{A1.13}$$

This is exactly the fluid equation. Just as in our Newtonian derivation, the fluid equation maintains energy conservation for the fluid as the Universe expands.

A1.3 Aside: Topology of the Universe

It is usually assumed that Universes with a flat or open geometry are infinite in extent, though the finite speed of light ensures we will never be able to prove this. However there is an alternative; the Universe may have a non-trivial **topology**. While geometry tells us the local shape of space or of space-time, topology describes the global properties. General relativity tells us that the properties of matter dictate the geometry, but says nothing about the topology.

The simplest type of topology is associated with identification of points in space. For example, if you take a sheet of paper and join two sides to form a cylinder, you have identified the points at these two edges. An ant can now walk around the cylinder forever, even though its extent in that direction is finite. If you can also bend the cylinder around to join its ends together, you can form a shape like a bagel, known formally as a torus. Like a sphere, we have a finite two-dimensional surface with no edge. However a torus and a sphere have a different global structure; for example if you draw any loop on a sphere it can always be continuously deformed so as to shrink away to nothing, while a loop on a torus which wraps round one of its principal directions can never be continuously removed; it is 'caught' around the hole. This difference is not a local geometric property of the surface, but is a global property of the entire surface and it indicates that the two surfaces have different topology.[1]

In fact a torus can be constructed with a flat geometry. The surface may look curved to you, but that curvature is just due to the way the surface has been represented in three-dimensional space. Any inhabitants restricted to the surface will find that the angles of any triangle always add up to 180°, and that circles have circumference equal to 2π times their radius. If you can only explore a small area, there is no way of telling whether you live on the surface of a torus or on a genuinely infinite plane.

It is possible for our Universe to have a non-trivial topology, so that for instance even if it has a flat geometry the volume might be finite and a traveller might return to their starting point in a finite time. If the scale of any topology is much larger than the observable

[1] Another example of identification is a video game where a character leaving one edge of the screen reappears on the other. If top and bottom are identified as well as the two edges, the game is actually taking place on the surface of a torus.

Universe then we have no way to detect it, but if it is smaller then there can be observable consequences. A flat geometry permits any scale of topology, and indeed even allows that scale to be different in different directions. If the topology scale were tiny, then light coming from a large distance would 'wrap' many times around the Universe before reaching us, and what we would see would be many repeats of the same set of galaxies in the same configuration. Such repeats are not seen, and so if there were topology its scale must be larger than any galaxy surveys yet carried out. This limit has been pushed out towards the size of the observable Universe by studies of distant quasars and particularly of the cosmic microwave background. No evidence of non-trivial topology has been seen, and upcoming cosmic microwave background observations should make definitive tests, with most cosmologists expecting topology to be ruled out as an interesting possibility. On the theoretical side, there are no well-motivated models predicting that there should be a non-trivial topology, and indeed discovery of topology of the Universe would be in conflict with standard inflationary cosmology models.

Non-trivial topology is also possible in the hyperbolic geometry. In principle this is a more interesting possibility, as there are only a certain set of possible sizes of topology (measured in comparison to the curvature scale), and in particular there is a smallest possible topology where the topology scale is about half the curvature scale. It is interesting to speculate that this might be favoured for some as-yet-unknown reason. However recent observations indicating that the Universe is very close to the flat geometry mean that even if the Universe is as far from flatness as allowed, and the topology scale the smallest possible, then the topology scale is still too large to be detectable.

Problems

A1.1. The spatial part of the Robertson–Walker metric is

$$ds_3^2 = a^2(t)\left[\frac{dr^2}{1-kr^2} + r^2\,d\theta^2 + r^2\,\sin^2\theta\,d\phi^2\right]$$

(a) For positive k, what is the allowed range of the r coordinate? Define a new coordinate by the transformation $r = (1/\sqrt{k})\sin\left(\sqrt{k}\,\xi\right)$, and rewrite the metric using it. Such coordinates are called hyperspherical coordinates. Explain why they show that the space has the geometry of a three-dimensional sphere.

(b) Find an analogous transformation for negative k. Use it to find the ratio of circumference to radius for a circle at coordinate distance $\xi = 10/\sqrt{|k|}$. Compare with the equivalent ratio for a flat geometry.

A1.2. What is the maximum possible physical separation in a closed Universe at the present epoch?

Advanced Topic 2

Classic Cosmology: Distances and Luminosities

Prerequisites: Chapters 1 to 7 and Advanced Topic 1.1

Observational cosmology considers how objects with given properties, such as luminosity and size, will appear to us. In particular, it is concerned with the dependence of that appearance on the cosmological model. The simplest manifestation is something we have already seen — the redshifting of light due to the expansion — but there are also important effects if the geometry is not the standard Euclidean one. The discussion will be focussed around the Robertson–Walker metric

$$ds^2 = -c^2dt^2 + a^2(t)\left[\frac{dr^2}{1-kr^2} + r^2\left(d\theta^2 + \sin^2\theta\, d\phi^2\right)\right], \qquad \text{(A2.1)}$$

but whereas in Advanced Topic 1.1 we were primarily concerned with the geometrical interpretation of the spatial part alone, in considering light propagation we require the full space-time.

A2.1 Light propagation and redshift

In Section 5.2 we derived the redshift of photon wavelengths in a rather heuristic manner. In this section, a rigorous general relativistic interpretation will be given.

The key property of light propagation is that it obeys

$$ds = 0. \qquad \text{(A2.2)}$$

That is to say, a light ray travels no distance at all in space-time. At a given time all points in space are equivalent, so for simplicity we can consider a light ray to propagate radially from $r = 0$ to $r = r_0$, giving $d\theta = d\phi = 0$. Remember that the spatial coordinates in the metric are *comoving*, so galaxies remain at fixed coordinates; the expansion is entirely taken care of by the scale factor $a(t)$.

Setting $ds = 0$ for our radial light ray tells us that

$$\frac{c\,dt}{a(t)} = \frac{dr}{\sqrt{1-kr^2}}\,. \tag{A2.3}$$

To find the total time the ray takes to get from $r = 0$ to $r = r_0$, we simply integrate this giving

$$\int_{t_e}^{t_r} \frac{c\,dt}{a(t)} = \int_0^{r_0} \frac{dr}{\sqrt{1-kr^2}}\,, \tag{A2.4}$$

where 'e' stands for emission and 'r' for reception.

Now consider a light ray emitted a short time interval later, so the emission time is $t_e + dt_e$. The galaxies are still at the same coordinates, so we can get the time of reception, $t_r + dt_r$, from a similar integral

$$\int_{t_e+dt_e}^{t_r+dt_r} \frac{c\,dt}{a(t)} = \int_0^{r_0} \frac{dr}{\sqrt{1-kr^2}}\,. \tag{A2.5}$$

The right-hand sides of these equations are equal, so we can write

$$\int_{t_e}^{t_r} \frac{c\,dt}{a(t)} = \int_{t_e+dt_e}^{t_r+dt_r} \frac{c\,dt}{a(t)}\,. \tag{A2.6}$$

Remember that the integrals are just the areas under curves, in this case the curve of $c/a(t)$, which is a reducing function if the Universe is expanding. Figure A2.1 shows the integrand, and the previous equation tells us that the area between the dashed lines is equal to that between the dotted lines. As the central area is common to both integrals, this implies that the two narrow slices at the edges must have the same area[1]

$$\int_{t_e}^{t_e+dt_e} \frac{c\,dt}{a(t)} = \int_{t_r}^{t_r+dt_r} \frac{c\,dt}{a(t)}\,. \tag{A2.7}$$

Since the slices are narrow the area is just width times height, so

$$\frac{dt_r}{a(t_r)} = \frac{dt_e}{a(t_e)}\,. \tag{A2.8}$$

In an expanding Universe, $a(t_r) > a(t_e)$, so $dt_r > dt_e$. The time interval between the two rays increases as the Universe expands.

Now imagine that, instead of being two separate rays, they correspond to successive crests of a single wave. As the wavelength is proportional to the time between crests, $\lambda \propto dt \propto a(t)$, and so

$$\frac{\lambda_r}{\lambda_e} = \frac{a(t_r)}{a(t_e)}\,. \tag{A2.9}$$

[1] Of course you can get that just by rearranging the limits — you don't really need the graph.

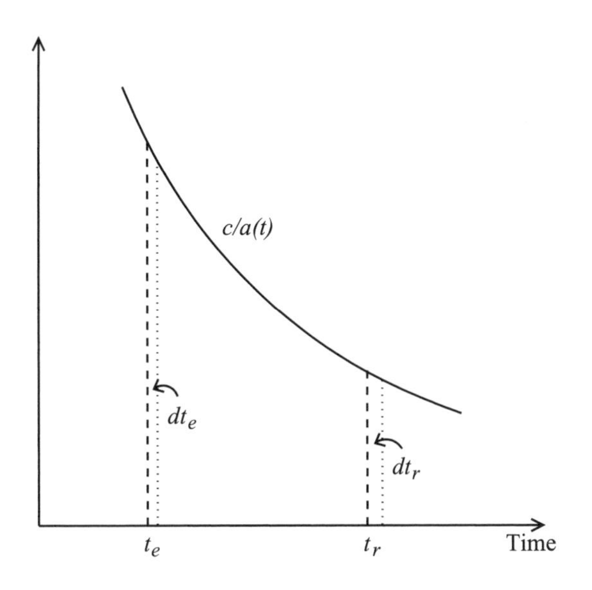

Figure A2.1 A graph of $c/a(t)$ illustrates how the redshift law can be derived.

This expression is exactly the one derived in Section 5.2, but now it applies for arbitrary separations and for any geometry of the Universe. The interpretation is that light is stretched as it travels across the Universe; for example if the light were intercepted by an intermediate observer comoving with the expansion, that observer would see the light with a wavelength intermediate to the original emitted and received wavelengths. Note that light emitted when $a(t) = 0$ would be infinitely redshifted.

The standard application of this expression is to light received by us, so that t_{r} is identified with t_0. We can then define the redshift z by

$$\frac{a(t_0)}{a(t_{\mathrm{e}})} \equiv 1 + z\,. \tag{A2.10}$$

It will be greater than zero in an expanding Universe, with $z \to \infty$ as we consider light emitted ever closer to the Big Bang itself.

It is quite common for astronomers to use the term 'redshift' to describe epochs of the Universe and to describe the distances to objects. For example, referring to the Universe at a redshift of z means the time when the Universe was $1/(1+z)$ of its present size. If an object is said to be at redshift z, that means that it is at a distance so that in the time its light has taken to reach us, it has redshifted by a factor $1 + z$. As I write, the most distant objects known are quasars at redshifts a little above 6, detected in the Sloan Digital Sky Survey, though new records are being set all the time. The most redshifted light we receive, however, is the cosmic microwave background radiation originating at $z \simeq 1000$.

A2.2 The observable Universe

Equipped with equation (A2.4), we can compute how far light could have travelled during the lifetime of the Universe. The distance is given by r_0 satisfying

$$\int_0^{r_0} \frac{dr}{\sqrt{1-kr^2}} = \int_0^{t_0} \frac{c\,dt}{a(t)} \,. \tag{A2.11}$$

Let's simplify by assuming a matter-dominated Universe with $k = \Lambda = 0$,[2] so that the relevant solution for the scale factor is $a(t) = (t/t_0)^{2/3}$ [equation (5.15)]. Then

$$\int_0^{r_0} dr = ct_0^{2/3} \int_0^{t_0} \frac{dt}{t^{2/3}} \quad \Longrightarrow \quad r_0 = 3ct_0 \,. \tag{A2.12}$$

Here r_0 is the coordinate distance, but in this example we have $a(t_0) = 1$ so physical distances and coordinate distances coincide.

There are two striking features of this result. The first is that at any given time it is finite; even though the solution for $a(t)$ we are using has $1/a(t) \to \infty$ at $t \to 0$, it can still be integrated. As nothing can travel faster than the speed of light, this means that even in principle we can only see a portion of the Universe, known as the **observable Universe**. However, if we have a different evolution of the scale factor at very early times this may lead to very different conclusions (e.g. see the discussion of inflation in Chapter 13). One way of circumventing this uncertainty is to instead define the observable Universe as the region that can be probed by electromagnetic radiation, noting that the Universe is opaque until the time of formation of the cosmic microwave background. Using this as the initial time gives a finite result which is independent of the (unknown) very early history of the Universe.

The second feature is that the distance the light has travelled is actually somewhat greater than the speed of light multiplied by the age of the Universe. This is because the Universe expands as the light crosses it; note that r_0 is the distance *as measured in the present Universe*. At early times when the Universe was smaller, it was easier for the light to make progress across it.

A2.3 Luminosity distance

The **luminosity distance** is a way of expressing the amount of light received from a distant object. Let us suppose we observe an object with a certain flux. The luminosity distance is the distance that the object appears to have, assuming the inverse square law for the reduction of light intensity with distance holds.

Let me stress right away that the luminosity distance is *not* the actual distance to the object, because in the real Universe the inverse square law does not hold. It is broken both because the geometry of the Universe need not be flat, and because the Universe is

[2] Using a more realistic cosmology changes the numbers, and can prevent an analytic derivation, but does not lead to qualitative changes.

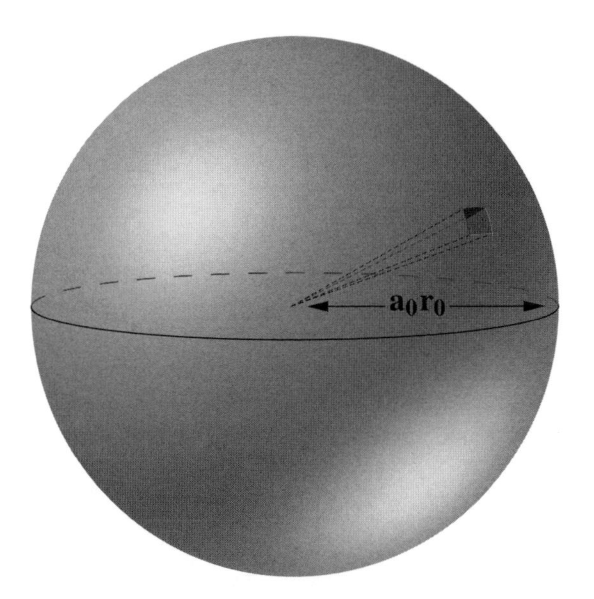

Figure A2.2 We receive light a distance $a_0 r_0$ from the source. The surface area of the sphere at that distance is $4\pi a_0^2 r_0^2$, and so our detector of unit area intercepts a fraction $1/4\pi a_0^2 r_0^2$ of the total light output $4\pi L$.

expanding. For generality, while in the following discussion it is presumed the object is observed at the present epoch, I will not set the present value of the scale factor a_0 to one.

We begin with definitions as follows. The luminosity L of an object is defined as the energy emitted per unit solid angle per second; since the total solid angle is 4π steradians, this equals the total power output divided by 4π. The radiation flux density S received by us is defined as the energy received per unit area per second. Then

$$d_{\rm lum}^2 \equiv \frac{L}{S}, \qquad \text{(A2.13)}$$

because L/S is the unit area per unit solid angle.

This is best visualized by placing the radiating object at the centre of a sphere, comoving radius r_0, with us holding our detector at the surface of the sphere, as shown in Figure A2.2. The physical radius of the sphere is $a_0 r_0$, and so its total surface area is $4\pi a_0^2 r_0^2$ (this fairly obvious answer can be verified explicitly by integrating the area element $r_0^2 \sin\theta \, d\theta \, d\phi$ from the metric over θ and ϕ). In this representation, the effect of the geometry is in the determination of r_0; it doesn't appear explicitly in the area.

If we were in a static space that would be the end of the story and the radiation flux received would simply be $S = L/a_0^2 r_0^2$, but we have to allow for the expansion of the Universe and how that affects the photons as they propagate from the source to the observer. There are actually two effects, which looks like double counting but is not:

- The individual photons lose energy $\propto (1+z)$, so have less energy when they arrive.

- The photons arrive less frequently $\propto (1+z)$.

Combining the two, the received flux is

$$S = \frac{L}{a_0^2 r_0^2 \left(1+z\right)^2}\,, \tag{A2.14}$$

and hence the luminosity distance is given by

$$d_{\rm lum} = a_0\, r_0 (1+z)\,. \tag{A2.15}$$

Distant objects appear to be further away than they really are because of the effect of redshift reducing their apparent luminosity. For example, consider a flat spatial geometry $k=0$. Then for a radial ray $ds = a(t)dr$, and so the physical distance to a source is given by integrating this at fixed time

$$d_{\rm phys} = a_0\, r_0 \tag{A2.16}$$

For nearby objects $z \ll 1$ and so $d_{\rm lum} \simeq d_{\rm phys}$, i.e. the objects really are just as far away as they look. But more distant objects appear further away ($d_{\rm lum} > d_{\rm phys}$) than they really are.

If the geometry is not flat, this gives an additional effect which can either enhance this trend (hyperbolic geometry) or oppose it (spherical geometry) — see Problem A2.2. Provided there is no cosmological constant, there are useful analytic forms for the luminosity distance as a function of redshift, related to an equation known as the Mattig equation, but once a cosmological constant is introduced calculations have to be done numerically. A detailed account can be found in Peacock's textbook (see Bibliography).

Before we can use the luminosity distance in practice, there is a problem to overcome. The luminosity we have described is the total luminosity of the source across all wavelengths (called the bolometric luminosity), but in practice a detector is sensitive only to a particular range of wavelengths. The redshifting of light means that the detector is seeing light emitted in a different part of the spectrum, as compared to nearby objects. If enough is known about the emission spectrum of the object, a correction can be applied to allow for this, which is known as the K-correction, though often its application is an uncertain business.

The luminosity distance depends on the cosmological model we have under discussion, and hence can be used to tell us which cosmological model describes our Universe. In particular, we can plot the luminosity distance against redshift for different cosmologies, as in Figure A2.3 (see Problem A2.4 to find out how to obtain these curves). Unfortunately, however, the observable quantity is the radiation flux density received from an object, and this can only be translated into a luminosity distance if the absolute luminosity of the object is known. There are no distant astronomical objects for which this is the case. This problem can however be circumvented if there are a population of objects at different distances which are believed to have the same luminosity; even if that luminosity is not known, it will appear merely as an overall scaling factor.

Such a population of objects is Type Ia supernovae. These are believed to be caused by the core collapse of white dwarf stars when they accrete material to take them over the Chandrasekhar limit. Accordingly, the progenitors of such supernovae are expected to be very similar, leading to supernovae of a characteristic brightness. This already gives a

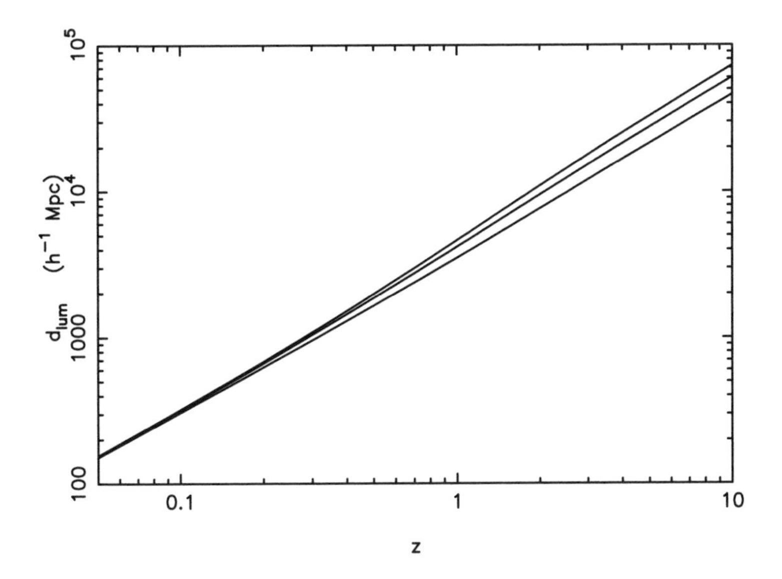

Figure A2.3 The luminosity distance as a function of redshift is plotted for three different spatially-flat cosmologies with a cosmological constant. From bottom to top, the lines are $\Omega_0 = 1$, 0.5 and 0.3 respectively. Notice how weak the dependence on cosmology is even to high redshift. It turns out that open Universe models with no cosmological constant have an even weaker dependence.

good standard candle, but it can be further improved as there is an observed correlation between the maximum absolute brightness of a supernova and the rate at which it brightens and fades (typically over several tens of days). And because a supernova at maximum brightness has a luminosity comparable to an entire galaxy, they can be seen at great distances.

Supernovae are rare events, but in the 1990s it became possible to systematically survey for distant supernovae by comparing telescope images containing large numbers of galaxies taken a few weeks apart. Two teams, the Supernova Cosmology Project and the High-z Supernova Search Team, were able to assemble samples containing tens of supernovae at redshifts approaching $z = 1$, and hence map the luminosity distance out to those redshifts.

The results delivered a major surprise to cosmologists. None of the usual cosmological models without a cosmological constant were able to explain the observed luminosity distance curve (usually called the apparent magnitude–redshift diagram). Figure A2.4 shows the allowed region of cosmological models in the Ω_0–Ω_Λ plane (as introduced in Section 7.3), with the contours indicating the allowed regions at different confidence levels. Results from the two collaborations are in excellent agreement, and in particular models with a flat spatial geometry agree with the supernova data only if $\Omega_0 \simeq 0.3$.

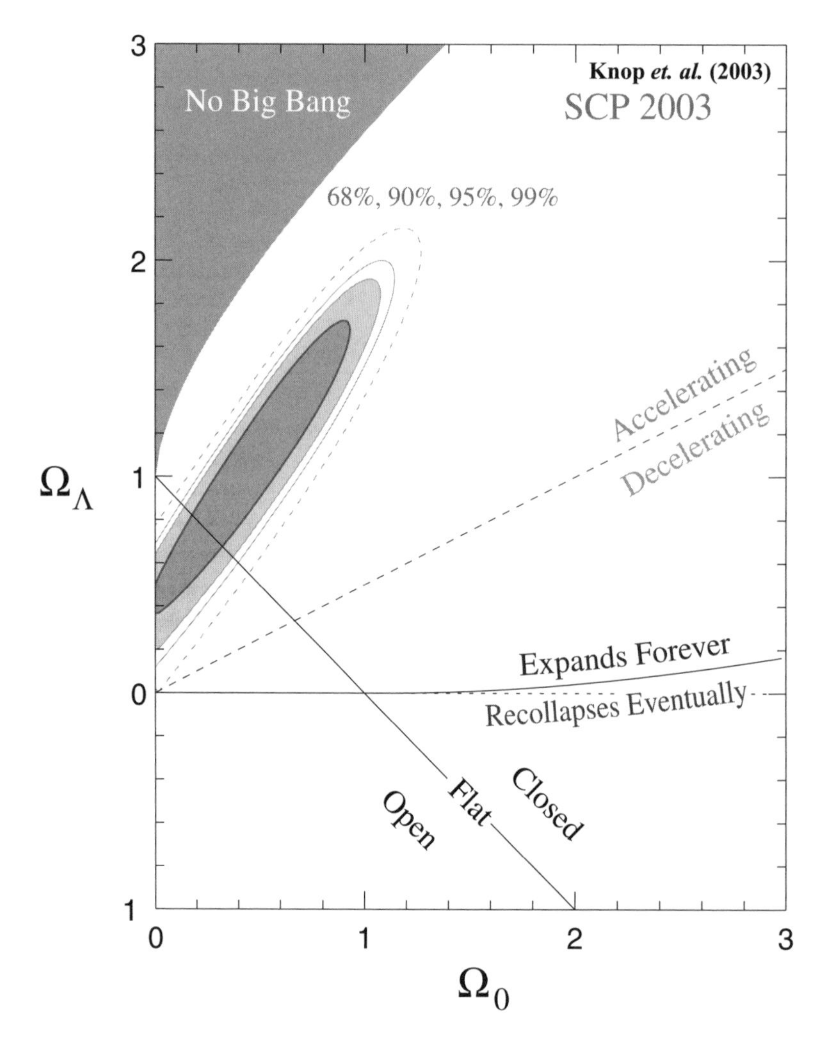

Figure A2.4 The contours show observational constraints on the supernova luminosity–redshift diagram obtained by the Supernova Cosmology Project, displayed in the Ω_0–Ω_Λ plane as introduced in Section 7.3. [From Knop et al., Astrophys. J **598**, 102 (2003), courtesy Supernova Cosmology Project.]

A2.4 Angular diameter distance

The **angular diameter distance** is a measure of how large objects appear to be. As with the luminosity distance, it is defined as the distance that an object of known physical extent appears to be at, under the assumption of Euclidean geometry. If we take the object to lie perpendicular to the line of sight and to have physical extent l, the angular diameter distance is therefore

$$d_{\rm diam} \equiv \frac{l}{\sin\theta} \simeq \frac{l}{\theta} \qquad \text{(A2.17)}$$

where the small-angle approximation used in the final expression is valid in almost any astronomical context.

To find an expression for this, it is most convenient this time to place ourselves at the origin, and the object at radial coordinate r_0. We need to use the metric at the time the light was emitted, t_e, and we align our 'rod' in the θ direction of the metric, equation (A2.1). The physical size l is measured using ds, now entirely in the θ direction, as

$$l = ds = r_0\, a(t_e)\, d\theta \tag{A2.18}$$

The light rays from each end of the rod propagate radially towards us, and so this angular extent is preserved even if the Universe is expanding. The angular size we perceive is

$$d\theta = \frac{l}{r_0\, a(t_e)} = \frac{l(1+z)}{a_0 r_0}\,, \tag{A2.19}$$

where the redshift term accounts for the evolution of the scale factor between emission and the present. Accordingly

$$d_{\rm diam} = \frac{a_0\, r_0}{1+z} \qquad \left[= \frac{d_{\rm lum}}{(1+z)^2} \right] . \tag{A2.20}$$

The angular diameter and luminosity distances therefore have similar forms, but have a different dependence on redshift.

As with the luminosity distance, for nearby objects the angular diameter distance closely matches the physical distance, so that objects appear smaller as they are put further away. However the angular diameter distance has a much more striking behaviour for distant objects. In discussing the observable Universe, we noted that even for distant objects $a_0\, r_0$ remains finite, but the light becomes infinitely redshifted. Hence $d_{\rm diam} \to 0$ as $z \to \infty$, meaning that distant objects appear to be nearby! Once objects are far enough away, moving them further actually makes their angular extent larger (though they do get fainter as according to the luminosity distance). In fact it is not hard to understand why, because the diameter distance refers to objects of fixed *physical* size l, so the earlier we are considering, the larger a comoving size they have. The angular diameter distance in three different cosmologies is shown in Figure A2.5.

In practice the Universe does not contain objects of a given fixed physical size back to arbitrarily early epochs. Nevertheless, objects of a given physical size appear smallest at a redshift $z \sim 1$ (with some dependence on the cosmological model chosen) and so one can hope to use distant objects to probe beyond the minimum angular size.

In a situation where we are observing distant objects at a high enough resolution that their angular extent is resolved (as is often the case for distant galaxies), the $(1+z)$ factors in both the luminosity and angular diameter distances can be relevant. The luminosity distance effect dims the radiation *and* the angular diameter distance effect means the light is spread over a larger angular area. This so-called surface brightness dimming is therefore a particularly strong function of redshift.

A key application of the angular diameter distance is in the study of features in the cosmic microwave background radiation, as described in Advanced Topic A5.4.

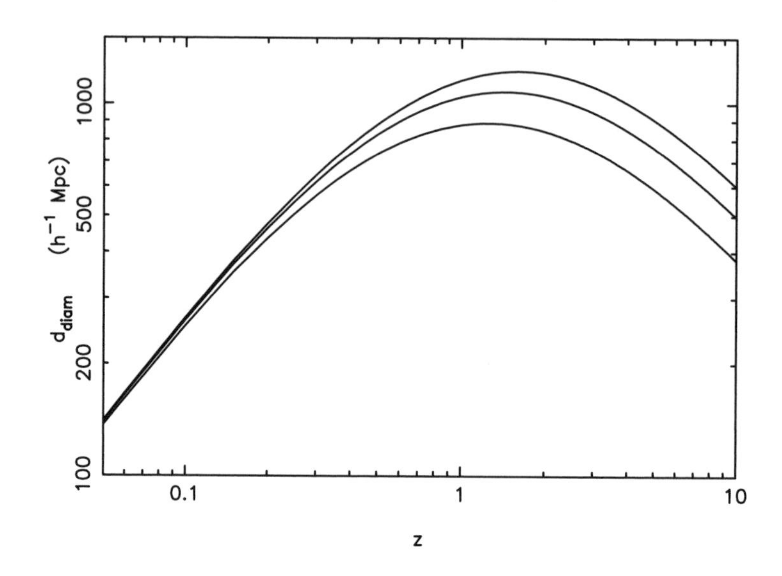

Figure A2.5 The angular diameter distance as a function of redshift is plotted for three different spatially-flat cosmologies with a cosmological constant. From bottom to top, the lines are $\Omega_0 = 1$, 0.5 and 0.3 respectively. For nearby objects d_{diam} and d_{lum} are very similar, but at large redshifts the angular diameter distance begins to decrease.

A2.5 Source counts

Another useful probe of cosmology is the source counts of objects (usually classes of galaxy in practical applications). Suppose sources are uniformly distributed in the Universe, with a number density $n(t) \propto 1/a^3$ that decreases with the expansion of the Universe. To compute the number of sources as a function of radius, we need the full volume element from the metric, which is the physical volume in an infinitesimal cube of sides dr, $d\theta$ and $d\phi$.

$$dV = \frac{a(t)\,dr}{\sqrt{1-kr^2}}\,a(t)\,r\,d\theta\,a(t)\,r\sin\theta\,d\phi\,. \tag{A2.21}$$

If we count sources per steradian, so that $\int \sin\theta\,d\theta\,d\phi = 1$, then the number of source dN in the volume element is

$$dN = \frac{n(t)a^3(t)r^2dr}{\sqrt{1-kr^2}} = \frac{n(t_0)a_0^3r^2dr}{\sqrt{1-kr^2}}\,, \tag{A2.22}$$

and the total number of sources per steradian out to distance r_0 is

$$N(r_0) = n(t_0)a_0^3\int_0^{r_0}\frac{r^2dr}{\sqrt{1-kr^2}}\,. \tag{A2.23}$$

For a useful application, one will have a limiting detectable flux in mind. To obtain the total number of sources, we use the luminosity distance to tell us how far away objects can be while still being bright enough to be seen, giving r_0. In principle source counts can be used to probe cosmological models, but in practice it is very difficult to untangle the effects of cosmology from evolution in the source population.

Problems

A2.1. A galaxy emits light of a particular wavelength. As the light travels, the expansion of Universe slows down and stops. Just after the Universe begins to recollapse, the light is received by an observer in another galaxy. Does the observer see the light redshifted or blueshifted?

A2.2. This question concerns the luminosity distance in a closed cosmological model, with metric given by equation (A2.1) with $k > 0$. The present physical distance from the origin to an object at radial coordinate r_0 is given by integrating ds at fixed time, i.e.

$$d_{\rm phys} = a_0 \int_0^{r_0} \frac{dr}{\sqrt{1 - kr^2}} \, .$$

Evaluate this (e.g. by finding a suitable change of variable) to show

$$d_{\rm phys} = \frac{a_0}{\sqrt{k}} \sin^{-1}\left(\sqrt{k} r_0\right) ,$$

and hence find an expression for $d_{\rm lum}$ in terms of $d_{\rm phys}$ and z. Show that $d_{\rm lum} \simeq d_{\rm phys}$ for nearby objects, and comment on the two effects which cause $d_{\rm lum}$ and $d_{\rm phys}$ to differ for distant objects.

A2.3. Throughout this question, assume a matter-dominated Universe with $k = 0$ and $\Lambda = 0$. By considering light emitted at time t_e (corresponding to a redshift z) and received at the present time t_0, show that the coordinate distance travelled by the light is given by

$$r_0 = 3ct_0 \left[1 - \frac{1}{\sqrt{1+z}}\right]$$

Derive a formula for the apparent angle subtended by an object of length l at redshift z. Find the behaviour in the limits of small and large z, and provide a physical explanation. Show that the object appears the smallest if it is located at redshift $z = 5/4$.

A2.4. Demonstrate that for spatially-flat matter-dominated cosmologies with a cosmological constant the Friedmann equation can be written as

$$H^2(z) = H_0^2\left[1 - \Omega_0 + \Omega_0(1+z)^3\right]$$

Use this to show that for spatially-flat cosmologies

$$r_0 = cH_0^{-1}\int_0^z \frac{dz}{\left[1 - \Omega_0 + \Omega_0(1+z)^3\right]^{1/2}}$$

Bearing in mind that $cH_0^{-1} = 3000\,h^{-1}\,\mathrm{Mpc}$, derive formulae for the luminosity and angular diameter distances as a function of redshift for the special case $\Omega_0 = 1$. If you are feeling adventurous, solve this equation numerically to obtain curves for $\Omega_0 = 0.3$ as shown in Figures A2.3 and A2.5.

A2.5. This question concerns what are called 'Euclidean number counts', meaning source counts in the limits where geometry and expansion can both be ignored so that the geometry is approximated as Euclidean. Evolution of the source population is also ignored.

Consider a population of sources with the same fixed luminosity, distributed throughout space with uniform number density. Determine how the number of sources seen above a given flux density limit S scales with S, and show that this scaling relation is unchanged if we have populations of sources with different absolute luminosities. Use this to argue that any realistic survey of objects is likely to be dominated by sources close to the flux limit for detection.

Advanced Topic 3

Neutrino Cosmology

Prerequisites: Chapters 1 to 12

Evidence has mounted in recent years that neutrinos must possess a non-zero rest mass. Studies of neutrinos coming from the Sun, those interacting in the Earth's atmosphere, and those created on Earth via nuclear interactions, have all shown evidence that neutrinos possess the ability to change their type as they travel. This phenomenon is known as **neutrino oscillations**, whereby for instance an electron neutrino may temporarily become a muon neutrino before oscillating back to its original type. This phenomenon can be understood in particle physics models, but only those where the neutrino rest-mass is non-zero. The evidence is now sufficiently strong that a non-zero rest mass should be taken as the working hypothesis.

Despite that, it has yet to become clear whether cosmologists should routinely worry about neutrino masses in constructing their models, because it may well be that the neutrino mass is too small to have a significant impact. The aim of this chapter is to investigate some of the consequences of neutrino mass, and to assess the circumstances in which it can play an important role. A much more complete account of neutrino cosmology, including the possibility of decay of heavy neutrinos, can be found in the textbook by Kolb & Turner (see Bibliography).

A3.1 The massless case

In order to judge whether the neutrino mass is important or not, we first need to understand the massless case better. The purpose of this section is to derive equation (11.1) for the cosmological density of neutrinos, in order to study under what circumstances it holds.

The reason why we expect there to be a neutrino background is because in the early Universe the density would be high enough for neutrinos to interact, and they would be created by interactions such as

$$\begin{aligned} \mathrm{p}+\mathrm{e}^- &\longleftrightarrow \mathrm{n}+\nu_{\mathrm{e}} \\ \gamma+\gamma &\longleftrightarrow \nu_\mu+\bar{\nu}_\mu \end{aligned}$$

At sufficiently early times, these interactions will ensure that neutrinos are in thermal equilibrium with the other particle species, in particular the photons.

If neutrinos had identical properties to photons that would be the end of the story; as there are three types of neutrino and one type of photon we would simply predict $\Omega_\nu = 3\Omega_{\rm rad}$. However this simple estimate fails for two reasons; neutrinos are fermionic particles while photons are bosons, and neutrinos have different interaction properties.

It is fairly easy to account for the fermionic properties of neutrinos. The Fermi–Dirac distribution is very similar to the Bose–Einstein distribution, equation (2.7), but with the minus sign on the denominator replaced by a plus sign. Because of this, the occupation numbers of the states at a given temperature is smaller for fermions, though the difference is only significant for low frequencies. To figure out how much smaller, one has to do the integral analogous to equation (2.9). It turns out that it is smaller by a factor $7/8$.

Much less trivial is accounting for the difference between photon and neutrino properties. In Chapter 10 we learnt that photons cease interaction (known as **decoupling**) at $T \simeq 3000\,\mathrm{K}$, but neutrinos interact much more weakly and hence decouple at a much higher temperature. The decoupling time can be estimated by comparing the neutrino interaction time with the expansion rate; if the former dominates then thermal equilibrium is maintained, but if the latter then the reactions are too slow to maintain equilibrium and can be considered negligible. The weak interaction cross-section gives the relevant interaction rate, and it can be shown (see Problem A3.1) that the interaction rate exceeds the expansion rate for $k_{\rm B}T > 1\,\mathrm{MeV}$; once the Universe falls below this temperature the neutrinos cease interacting.

The significance of this is that this temperature is above the energy at which electrons and positrons are in thermal equilibrium with the photons; as the electron rest mass–energy is $0.511\,\mathrm{MeV}$, provided the typical photon energy is above this electron–positron pairs are readily created (and destroyed) by the reaction

$$\gamma + \gamma \longleftrightarrow \mathrm{e}^+ + \mathrm{e}^-$$

and so at $k_{\rm B}T \simeq 1\,\mathrm{MeV}$ we expect electrons and positrons to have similar number density to photons.[1] Once the temperature falls further, the photons no longer have the energy to create the pairs and the reaction above proceeds only in the leftwards direction, with electron–positron annihilation leading to the creation of extra photons. The corresponding cross-section for electrons and positrons to annihilate into neutrinos is vastly smaller, so the annihilations serve to create extra photons but not neutrinos, as there is no mechanism to transfer the excess energy into the neutrinos.

Once the annihilations have created these new photons, the photons rapidly thermalize amongst themselves, boosting their temperature relative to that of the neutrinos. It turns out that the decays take place at constant entropy, and this can be used to show that the temperature increases by the curious factor of $\sqrt[3]{11/4}$ (see Problem A3.2). We know the present photon temperature is $2.725\,\mathrm{K}$, so the present neutrino temperature is predicted to be

$$T_\nu = \sqrt[3]{\frac{4}{11}}\,T = 1.95\ \mathrm{Kelvin}\,. \tag{A3.1}$$

[1] You might worry that there isn't much difference between the thermal energy at neutrino decoupling and the electron mass–energy, but detailed calculations show that the difference is enough that the events can be considered to take place sequentially.

Putting all these pieces together, and remembering that the energy density goes as the fourth power of the temperature, we conclude that

$$\Omega_\nu = 3 \times \frac{7}{8} \times \left(\frac{4}{11}\right)^{4/3} \Omega_{\rm rad} = 0.68\,\Omega_{\rm rad} = 1.68 \times 10^{-5} h^{-2} \,. \qquad \text{(A3.2)}$$

The validity of this expression requires that the neutrinos act as relativistic species, i.e. that their rest mass–energy is negligible compared to their kinetic energy. The kinetic energy per particle is about $3k_{\rm B}T_\nu \simeq 5 \times 10^{-4}\,{\rm eV}$. The above calculation of the neutrino energy density is therefore valid only if the masses of all three neutrino species are less than this. If the masses exceed this, the neutrinos would be non-relativistic by the present and this would need to be accounted for.

A3.2 Massive neutrinos

While a neutrino mass–energy greater than $5 \times 10^{-4}\,{\rm eV}$ would have an important effect at the present epoch, it would have to be much higher in the early history of the Universe for it to play an important role as the neutrino thermal energy was much higher then. In particular, we can distinguish two cases depending on whether or not the mass–energy is negligible at neutrino decoupling.

A3.2.1 Light neutrinos

At neutrino decoupling, the thermal energy is $k_{\rm B}T \simeq 1\,{\rm MeV}$. If the neutrino mass–energy is much less than this, it will be unimportant at the decoupling era, which is when the number density of neutrinos is determined. I will refer to this regime as **light neutrinos**.

In the case of light neutrinos the formula for the cosmological density of neutrinos is easily derived. The number of neutrinos is just the same as in the massless case, but instead of their kinetic energy $5 \times 10^{-4}\,{\rm eV}$, their mass–energy is now dominated by their rest mass–energy $m_\nu c^2$. If we consider just one species of massive neutrino, the corresponding energy density would therefore be

$$\Omega_\nu = \frac{1.68 \times 10^{-5} h^{-2}}{3} \frac{m_\nu c^2}{5 \times 10^{-4}\,{\rm eV}} = \frac{m_\nu c^2}{94\,h^2\,{\rm eV}} \,. \qquad \text{(A3.3)}$$

For the more likely case that all neutrinos have a mass, this can be written

$$\Omega_\nu = \frac{\sum m_\nu c^2}{94\,h^2\,{\rm eV}} \,, \qquad \text{(A3.4)}$$

where the sum is over the neutrino types with $m_\nu c^2 \ll 1\,{\rm MeV}$.

We see that a light neutrino species could readily provide the observed dark matter density $\Omega_{\rm dm} \simeq 0.3$. Taking $h = 0.72$, it requires a neutrino of mass–energy $m_\nu c^2 \simeq 14\,{\rm eV}$. This is above current experimental limits for the electron neutrino, but acceptable for the other two species. However, in fact such a neutrino is not thought to be a good dark matter candidate, because it is relativistic until fairly late in the Universe's evolution (see Problem A3.3) which prevents the formation of galaxies.

Equation (A3.4) is a powerful constraint on neutrino properties. As we do not believe the matter density exceeds the critical density,[2] *stable* neutrinos cannot have a mass in the range from $90\,h^2\,\mathrm{eV}$ all the way up to the $1\,\mathrm{MeV}$ for which the calculation is valid (the next subsection will explore higher masses). Such neutrinos might be permitted if they proved to be unstable, though that would depend in detail on the nature of the decay products.

A3.2.2 Heavy neutrinos

If the neutrino masses exceed 1 MeV, the calculation of the neutrino density needs further modification, and I will refer to this limit as **heavy neutrinos**. In this case, at neutrino decoupling the neutrino mass–energy is already higher than the thermal energy. In this regime the number density of neutrinos is suppressed, the most important term being an exponential (Boltzmann) suppression factor $\exp\left(-m_\nu c^2/k_\mathrm{B}T\right)$. The higher the neutrino mass, the more potent this suppression, and hence the predicted neutrino mass density begins to fall as the exponential suppression of the number density overcomes the extra mass-per-particle. A detailed calculation shows that once $m_\nu c^2$ reaches around 1 GeV, the predicted neutrino density has once more fallen to $\Omega_\nu \sim 1$; this analysis therefore extends the excluded mass–energy range for neutrinos discussed in the previous subsection up to 1 GeV. Heavy neutrinos with $m_\nu c^2 \simeq 1\,\mathrm{GeV}$ are therefore another candidate to be the dark matter in the Universe, but this time they become non-relativistic extremely early and are a cold dark matter candidate. This is very much what we would like for successful structure formation, but unfortunately laboratory limits on all three known neutrino species are well below 1 GeV. Accordingly, only some new type of neutrino, perhaps with unconventional interactions, could fulfil that role, which is an unattractive proposition.

A3.3 Neutrinos and structure formation

The cross-section for neutrino interactions with normal matter depends on the neutrino momentum (see Problem A3.1), and the very low momenta predicted for the cosmic neutrinos means they cannot be detected by any existing or planned detector. Nevertheless, it should be possible to verify the existence of cosmic neutrinos indirectly via their effect on structure formation. Between neutrino decoupling and photon decoupling the two species have very different properties, the former travelling freely and the latter still strongly interacting with baryonic matter.

In the case of light neutrinos massive enough to contribute significantly to the dark matter density, there are already strong limits. These neutrinos correspond to hot dark matter, meaning particles which, though non-relativistic now, travelled a significant distance while relativistic (see Problem A3.3). This opposes the formation of structure and

[2]Although commonly used by cosmologists, this phrasing is rather sloppy; we have already seen that if Ω_0 is initially equal to one then it remains so for all time, regardless of how many neutrinos might be formed. A more technically correct version of this argument would compute the ratio of neutrino density to photon density, and impose a limit from combining the requirements that the total density must not exceed one and that the photon density has its observed value.

can prevent galaxies from forming, and pure hot dark matter is strongly excluded by observations.

Relativistic neutrinos also have important effects at early times. Most important is on nucleosynthesis, where the presence of a neutrino background appears essential to obtain the right element abundances. However there are also predicted effects on structure formation. The epoch of matter–radiation equality computed in equation (11.5) would be different if the neutrinos were omitted, and it turns out that this epoch gives a characteristic scale in the clustering of galaxies. The existence of the neutrino background also plays an important role in predictions of structures in the cosmic microwave background, with different results obtained if the neutrinos are not present. As I write, these observations are not of high enough quality to unambiguously confirm the existence of the cosmic neutrino background at the expected level, but they may well soon be.

Problems

A3.1. The decoupling temperature of neutrinos can be estimated by comparing the typical interaction rate with the expansion rate H of the Universe. The cross-section for weak interactions depends on momentum (and hence temperature), and is given by $\sigma \simeq G_{\rm F}^2 p^2$ where p is the momentum and the Fermi constant $G_{\rm F} = 1.17 \times 10^{-5}\,{\rm GeV}^{-2}$. [For simplicity I have set $c = \hbar = 1$ in this question; if you want to include them you need to multiply by a term $(\hbar c)^{-4}$ on the right-hand side.] Assuming the neutrinos are highly relativistic, write this in terms of the temperature, taking the characteristic energy as $k_{\rm B}T$.

With $c = \hbar = 1$, the number density of relativistic species is $n \simeq k_{\rm B}^3 T^3$ (for the temperatures we are interested in the coefficient happens to be close to unity) and the Friedmann equation can be approximated as

$$H^2 = \frac{k_{\rm B}^4 T^4}{(10^{19}\,{\rm GeV})^2}\,.$$

Obtain an expression for the interaction rate per neutrino, Γ, and show that

$$\frac{\Gamma}{H} \simeq \left(\frac{k_{\rm B}T}{1\,{\rm MeV}}\right)^3 .$$

This confirms that the neutrino decoupling temperature is around 1 MeV.

A3.2. The entropy density of a sea of relativistic particles at temperature T is given by

$$s = \frac{2\pi^2}{45}\, g_*\, T^3 \,,$$

where g_* is the number of particle degrees of freedom and again fundamental constants have been set to one. Fermions count $7/8$ per degree of freedom towards this sum, and bosons 1. Photons have two degrees of freedom (the polarization states), and each of the electron and positron has two states (spin up and spin down). If the epoch of electron–positron annihilation occurs at constant entropy and produces only photons, demonstrate that the photon temperature is raised relative to the neutrino temperature by a factor $\sqrt[3]{11/4}$.

A3.3. By considering the ratio of the neutrino thermal energy $3k_{\mathrm{B}}T$ to its mass–energy, derive an approximate formula for the redshift at which massive neutrinos first become non-relativistic. Evaluate this redshift for the case of a neutrino hot dark matter candidate with mass–energy $m_\nu = 10\,\mathrm{eV}$. Using equation (11.12), estimate the distance (in *comoving* megaparsecs) that such neutrinos travel while relativistic.

Advanced Topic 4

Baryogenesis

Prerequisites: Chapters 1 to 12

Chapter 12 described the theory of cosmic nucleosynthesis, and demonstrated that good agreement with the observed light element abundances is only achieved if the baryon density satisfies the tight constraint

$$0.016 \leq \Omega_{\mathrm{B}} h^2 \leq 0.024 \,. \tag{A4.1}$$

Since we have such an accurate measure of the observed baryon density in the Universe, it would be nice to have a theory explaining its value, in the same way that the theory of nucleosynthesis explains the abundances of the light elements. Such theories are known as **baryogenesis**, but unfortunately at present are highly speculative and have no pretence of matching the observational accuracy. Rather, the current goal is to obtain an order-of-magnitude understanding of the baryon-to-photon ratio of 10^{-9}, and even that has yet to be achieved. It seems undesirable to assume that the Universe began with the baryon asymmetry already in place, so the currently-favoured models assume that there are processes which preferentially create matter rather than anti-matter, and try to exploit them.

In order to generate a baryon asymmetry, there are three conditions which must be satisfied, known as the **Sakharov conditions** after Andrei Sakharov who first formulated them in 1967. They are

1. Baryon number violation.

2. C and CP violation.

3. Departure from thermal equilibrium.

Clearly baryon number violation is necessary to generate a baryon asymmetry. Interactions in the Standard Model of particle physics conserve baryon number, meaning that the total number of baryons at the end of any interaction is the same as at the start. New types of interactions are needed to satisfy the first Sakharov condition; for example Grand Unified Theories seeking to merge the fundamental forces of nature typically permit baryon number violating interactions.

C and CP violation refers to two symmetries typically obeyed by particle interactions — that interaction rates are unchanged if one switches the charge (C) or parity (P) of the particles, or both (CP). For our discussion, the desired property is that anti-particles don't behave in precisely the same way as particles. If they did, then if we start with an equal mix of particles and anti-particles, any baryon number that might be generated through interactions of the particles will be exactly cancelled out by the equivalent interactions of the anti-particles. CP violation is observed in interactions of particles called neutral K-mesons, though at a very low level and without the presence of baryon number violation. Such violation, at a much larger level, would be needed in any interactions able to generate the baryon number.

Finally, thermal equilibrium is characterized by all interactions proceeding at the same rate in both the forward and backward directions. If the Universe stayed in thermal equilibrium, it wouldn't matter whether any interactions might generate a baryon number, because the reverse reactions would cancel it out. Departure from thermal equilibrium permits reactions to run preferentially in one direction. In cosmology, we expect the cooling due to the expansion of the Universe to lead to occasional departures from thermal equilibrium, as the available energy becomes too small to create massive particles, existing ones of which then subsequently decay. A typical baryogenesis scenario might therefore exploit a massive particle whose decays violate both baryon number and C/CP symmetry.

Although no established models exist, the overall picture of what is required is quite simple. Usually, the matter–anti-matter asymmetry is thought to have been created very early in the history of the Universe. When the mean photon energy was much higher than the baryon rest mass, $k_{\mathrm{B}}T \gg m_{\mathrm{p}}c^2$, it was possible to create baryons and anti-baryons in thermal equilibrium, by reactions such as

$$\gamma + \gamma \longleftrightarrow \mathrm{p} + \bar{\mathrm{p}}\,, \tag{A4.2}$$

where $\bar{\mathrm{p}}$ is an anti-proton. At these times one expects as many protons and anti-protons as photons of light. This is an ideal time to set about making a matter–anti-matter asymmetry; all one has to do is create one extra proton for every billion which exist, while leaving the anti-protons untouched. At this point the story becomes rather weak, because there is no established theory of how this might happen, but let's suppose that there exists a heavy particle, which we will call **X**, with suitable baryon number violating decays which is also produced in the thermal bath. It and its anti-particle should initially also be present in the same number as protons. As the Universe cools, there is insufficient energy to generate these heavy particles via interactions, and those particles in existence begin to decay, generating the baryon number. This process need only have an efficiency such that for every billion **X** and $\bar{\mathbf{X}}$ particles that decay, a single baryon is preferentially created.

Having created this minor imbalance and reached a stage where baryon number is conserved, we simply wait for the Universe to cool, and once $k_{\mathrm{B}}T \ll m_{\mathrm{p}}c^2$ the protons and anti-protons will annihilate. There will be too little energy to create new ones. Out of each one billion and one protons, one billion of them annihilate with the one billion anti-protons, and the remaining one is left over. This will give the required baryon density, as we only need one proton for every billion or so photons of light. If this picture, shown schematically in Figure A4.1, turns out to be true, then all the baryons we see, including those we ourselves are made of, have their origin in the small initial excess.

While the picture described above is the simplest, there are other ideas for generating the baryon asymmetry. One of the most important is **electro-weak baryogenesis**. While particle interactions involving the weak force, part of the Standard Model of par-

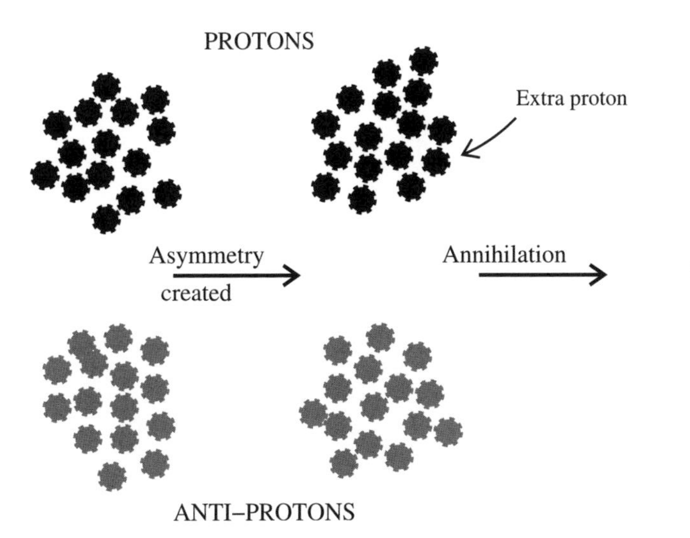

Figure A4.1 The favourite way to make a matter–anti-matter asymmetry is to do so very early, when the Universe was full of baryons and anti-baryons, by making a small excess of baryons. (I've contented myself with drawing fifteen rather than a billion!) Later, when the baryons and anti-baryons annihilate, the small excess is left over.

ticle physics, conserve baryon number, it was discovered by Gerard t'Hooft that a more complicated type of interaction (catch-phrases *non-perturbative* or *sphaleron*), which can be thought of as a type of many-particle interaction, actually does violate baryon number, opening the possibility of baryogenesis without considering new interactions. In the cool conditions of the present Universe sphaleron interactions are negligibly rare and so baryon number is indeed conserved, but in the early Universe they may be frequent. At very high energies these interactions have the tendency to try and suppress any pre-existing baryon asymmetry (as such they act against the type of scenario outlined above), while in the process of going out of equilibrium they may be able to generate an asymmetry. Unfortunately current calculations indicate that the sphaleron interactions are too inefficient to give the observed baryon number, leaving such studies at an impasse.

In summary, the accurate observation of the baryon density of the Universe presently lacks a satisfying explanation in terms of fundamental physics. It may be that new theoretical ideas are needed before progress can be made in this direction.

Advanced Topic 5
Structures in the Universe

Prerequisites: Chapters 1 to 13

This book is about the Hot Big Bang model and its successes in describing the Universe as a whole. The basic precept has been the cosmological principle, requiring that the Universe be homogeneous and isotropic, and we have seen how this leads to an explanation for the cosmic microwave background and the light element abundances. However, although the cosmological principle is valid for studying the Universe as a whole, we know that it doesn't hold perfectly. The nearby Universe is highly inhomogeneous, being made up of stars and planets and galaxies rather than the smoothly-distributed fluid of mass density ρ that we've considered so far. Attempting to explain these observed structures is the most active research area in modern cosmology, and no cosmology textbook would be complete without making some mention of it. However, this book is not the place to develop the detailed mathematics of structure formation, and I will mostly keep things at a descriptive level. The details can be found in the more advanced textbooks listed in the Bibliography.

A5.1 The observed structures

The brief observational overview of Chapter 2 showed you some of the observed structures in the Universe. The fundamental building blocks of cosmology are galaxies. These come in a wide variety of types, some spiral, some elliptical and some with irregular shapes. They also have a wide range of masses, from dwarf galaxies of only a million solar masses up to giant galaxies lurking in the centre of galaxy clusters, which might be ten times more massive than our own. As we saw in Figure 2.2, galaxies are not distributed randomly in space, but rather show strong clustering with the galaxies lining up in filaments and walls, with large voids in between. You are far more likely to find a galaxy near another galaxy than at a randomly-selected location.

Places where galaxies are grouped together so closely that they are held together by their mutual gravitational attraction are called galaxy clusters. An example is the Coma cluster mentioned in Chapter 2. The individual galaxies are on orbits, often highly eccentric ones, around the centre of mass, and the largest galaxy clusters contain thousands of galaxies. Even galaxy clusters are themselves clustered, again meaning that if you want to find a galaxy cluster, the best place to look is near another one.

The distribution of galaxies in the Universe has been studied for several decades now. A much more recent, and rapidly advancing, field of observational cosmology is the study of irregularities in the cosmic microwave background radiation. Because we study the radiation coming from different directions, these are known as **anisotropies**. As the cosmological principle is not exact, it had long been expected that anisotropies must exist in the microwave background radiation at some level. In practice, they proved extremely hard to detect, and it was not until 1992 that they were measured by the DMR (Differential Microwave Radiometer) experiment on the COBE satellite. The typical difference in temperature, ΔT, if you look in two different directions turns out to be only a few tens of microKelvin (one Kelvin equals one million microKelvin). Remembering that the average temperature is $2.725\,\mathrm{K}$, that means that the fractional irregularity in temperature is

$$\frac{\Delta T}{T} \sim 10^{-5}\,. \tag{A5.1}$$

If a swimming pool were to be so smooth, the largest ripples could only be one hundredth of a millimeter high!

The microwave background has been travelling towards us uninterrupted since decoupling, when the Universe was only $350\,000\,\mathrm{yrs}$ old. Its anisotropies carry a record of the state of the Universe at that time, and are telling us that the Universe was much more homogeneous then than it is now. Since COBE, many other experiments have been able to make measurements of the anisotropies, and their study is becoming a mature science.

The nearby galaxy distribution shows us the present Universe, while the microwave background probes a very early stage of its history. It is only recently that significant inroads have been made in understanding what happened between those epochs, by studying galaxies with high redshifts. Recall the relationship between redshift and scale factor, equation (5.10)

$$1 + z = \frac{a_{\mathrm{obs}}}{a_{\mathrm{em}}}\,, \tag{A5.2}$$

where the right-hand side has the scale factor at the time of observation (meaning now) and the time of emission of the light. If the light from a galaxy indicates a high redshift, say of 3 or 4, then the light must have been emitted when the Universe was a small fraction of its present size. If we specialize to a critical-density matter-dominated Universe, so that $a = (t/t_0)^{2/3}$, then the time the light was emitted was

$$\frac{t}{t_0} = \frac{1}{(1+z)^{3/2}}\,, \tag{A5.3}$$

where t_0 is the present age of the Universe. So an object with redshift $z = 4$, for example, is being seen when the Universe was only about $1/11$-th of its present age, i.e. around one billion years old. Since its light has been travelling towards us since then, it has come from a considerable distance away, most of the way across the observable Universe.

Powerful telescopes, such as the Hubble Space Telescope and the Keck Telescope in Hawaii, are capable of studying galaxies at these redshifts. In fact, the furthest galaxies seen have redshifts above five. Samples of galaxies with redshifts around three have become large enough that the galaxies' clustering can be studied, and compared with the clustering of galaxies in our local neighbourhood.

All of this reconfirms our view of the Universe as an evolving place. At decoupling the

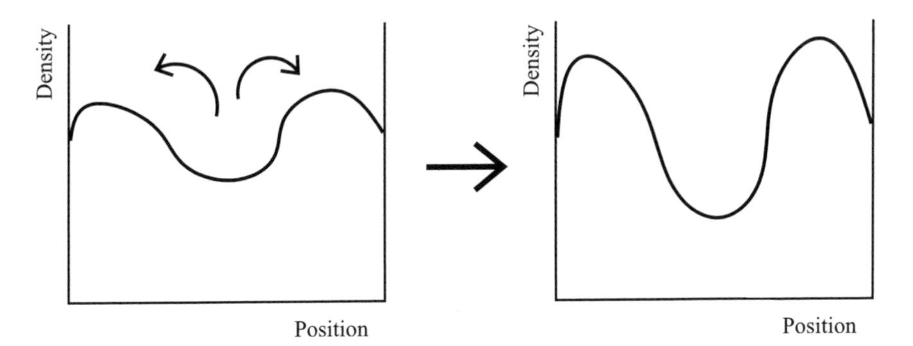

Figure A5.1 Gravity pulls material towards the denser regions, enhancing any initial irregularities.

irregularities were very small, and so the microwave background is very close to isotropy. The high-redshift galaxies seen by the Hubble Space Telescope appear different from those around us, being more likely to have irregular shapes and to be involved in interactions with other galaxies, and so the galaxy population has evolved since then. Presumably even the present state of the Universe is a transient one, and things will also look different in the distant future.

The various forms of structure in the Universe are often collectively referred to as **large-scale structure**.

A5.2 Gravitational instability

The key idea in explaining the way in which structures evolve in the Universe is **gravitational instability**. If material is to be brought together to form structures then a long-range force is required, and gravity is the only known possibility. (Although electromagnetism is a long-range force, charge neutrality demands that its influence is unimportant on large scales.) The basic picture is as follows.

Suppose that at some initial time, say decoupling, there are small irregularities in the distribution of matter. Those regions with more matter will exert a greater gravitational force on their neighbouring regions, and hence tend to draw surrounding material in. This extra material makes them even denser than before, increasing their gravitational attraction and further enhancing their pull on their neighbours, shown in Figure A5.1. An irregular distribution of matter is therefore unstable under the influence of gravity, becoming more and more irregular as time goes by.

This instability is exactly what's needed to explain the observation that the Universe is much more irregular now than at decoupling, and gravitational instability is almost universally accepted to be the primary influence leading to the formation of structures in the Universe. It's an appealingly simple picture, rather spoiled in real life by the fact that while gravity may have the lead role, numerous other processes also have a part to play and things become quite complicated. For example, we know that radiation has pressure proportional to its density, and during structure formation the irregularities create pressure gradients which lead to forces opposing the gravitational collapse. We know that neutrinos move relativistically and do not interact with other material, and so they are able to escape from structures as they form. And once structure formation begins, the complex astrophysics of stars, especially supernovae, can inject energy back into the intergalactic regions and influence regions yet to complete their gravitational collapse.

The formation of structure is a massive subject area, and in this book I will consider only one facet of it — the use of structure formation studies to constrain the cosmological model. Clearly the cosmological setting, for example the material composition of the Universe and its rate of expansion, will influence the way gravitational instability develops, and so accurate studies of structure can be used to constrain cosmological models. This endeavour is usually called **parameter estimation**, the presumption being that the basic cosmological model has been established, leaving us the task of finding the actual values of the various parameters, such as H_0 and Ω_Λ, which describe our own Universe. Parameter estimation now typically involves around ten parameters, some describing the present dynamical state and matter content of the Universe and others describing the nature of the primordial irregularities that initiate structure formation (see Advanced Topic 5.5). Different kinds of observations tend to be sensitive to different subsets of the complete parameter set, so a fully comprehensive parameter estimation study will consider several different types of data.

A5.3 The clustering of galaxies

The clustering of galaxies is the oldest topic in structure formation, and the phrase 'large-scale structure' is still sometimes used to refer to it alone. Early studies, culminating in the CfA survey (Figure 2.2) in the mid nineteen-eighties, were able to establish the three-dimensional positions of several thousands of galaxies, clearly showing substantial clustering. Spectra of each galaxy were obtained in order to determine their redshift, which using the Hubble law gives their distance from us (at least if the peculiar velocity of the galaxy, which cannot easily be independently measured, is small compared to their recession velocity).

Modern surveys are able to use multi-object spectrographs, where the light from several hundred galaxies can be collected simultaneously and fed to a spectrograph via fiber optic cables. These have revolutionized galaxy surveys, allowing them to include hundreds of thousands of galaxies. The 2dF galaxy redshift survey, shown in Figure A5.2, contains over 220,000 galaxies and was completed in mid 2002. An even larger effort, the Sloan Digital Sky Survey, reached over 650,000 galaxies by the end of its first phase in 2005 and is continuing through a second and third phase.

In order to analyze such large samples, the vast information needs to be condensed into statistical measures of clustering. The simplest are the two-point correlation function $\xi(r)$, which measures the likelihood of two galaxies having a given separation r, and the power spectrum $P(k)$ which decomposes the pattern into waves with wavenumber $k \equiv 2\pi/\lambda$ and gives the typical amplitude of those waves.

For the measurements to be useful, we must be able to predict those quantities for a given cosmological model, and determine how they vary with different choices of the cosmological parameters in order to determine which cosmologies best fit the data. This is a complex task requiring numerical solution of equations describing the evolution of structure, well beyond the scope of this book (see the Bibliography for a selection of advanced texts). During the initial stages of structure formation, where the irregularities are small

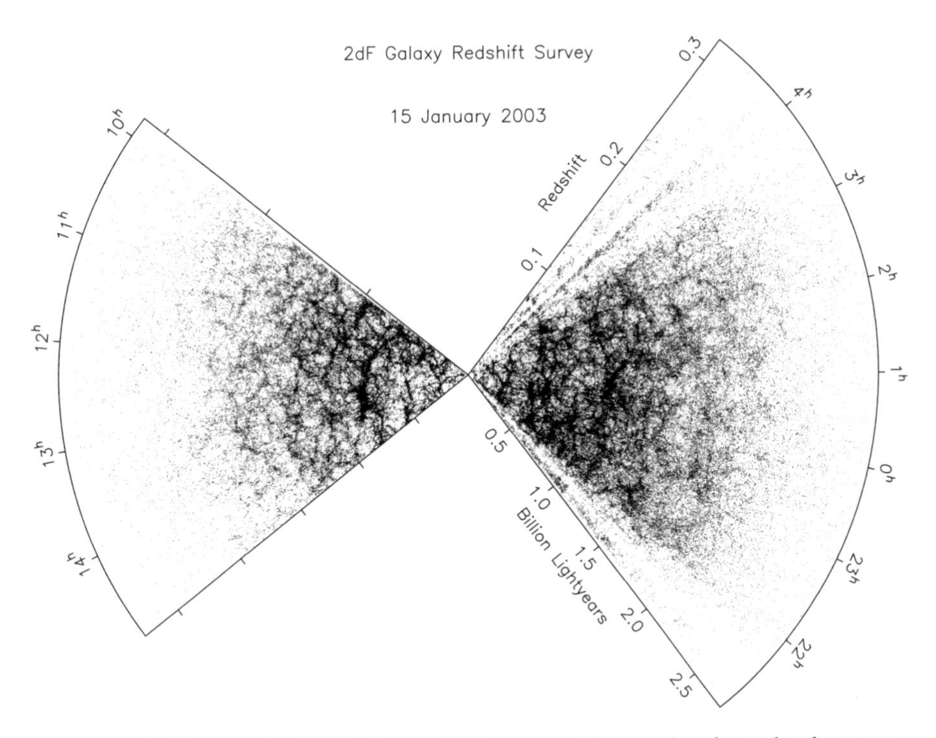

Figure A5.2 The completed 2dF galaxy redshift survey, with our galaxy located at the centre. The three-dimensional survey volume has been flattened to make this image, and the sudden angular variations indicate regions which were not surveyed. The number of galaxies is so large that projection effects make it difficult to see the structures. The radial distance indicates the redshift to the galaxy, which is independent of the underlying cosmological model, while the scale indicating distance makes particular assumptions about the cosmological model. For comparison, the CfA galaxy survey extends only to a redshift of around 0.02. [Figure courtesy Matthew Colless and the 2dFGRS Team.]

deviations from a homogeneous Universe, it is possible to follow their progress accurately, and the cosmology community has been greatly helped by researchers making computer codes to carry out these calculations publically available, such as the CMBFAST program of Seljak and Zaldarriaga. During the later stages, when gravitational collapse leads to the formation of individual galaxies and stars and supernovae begin to feed energy back into the intergalactic medium, the situation becomes highly complex, and even on state-of-the-art supercomputers it is possible to include only a subset of the physical processes that might be relevant.

The best understanding is of the distribution of the dark matter in the Universe; because it is the dominant form of material (apart from the cosmological constant which, being constant, is not able to form irregularities), and because it only interacts gravitationally, it is almost unaffected by other physical processes which might affect protons and neutrons. It is believed that we have a very good theoretical understanding of the dark matter distribution in the Universe, the only problem being that since it is dark we are unable to study it directly in order to see whether we are right! Galaxies presumably generally follow this dark matter distribution, since it is the gravitational attraction of the dark matter which causes galaxies to clump together, but the way in which they do this is likely to be complex — the jargon is that galaxies are **biassed** with respect to the dark matter. For instance, there is no good reason to think that a region with twice the average number of galaxies necessarily also has twice the amount of dark matter. It is hoped that in the near future measurements of the bending of light paths from distant sources caused by the presence of dark matter, known as gravitational lensing, will provide a more accurate way of measuring the dark matter distribution in order to test the theoretical predictions.

Despite these difficulties in matching theory to observation, it is clear that current cosmological models give an excellent description of the observations. On large scales of tens of megaparsecs and above, where the Universe is approaching homogeneity and accurate calculations can be made, the observed clustering can be accurately matched in cosmological models of the type discussed throughout this book. On shorter scales, it is much harder to extract information from observations which can be used to constrain cosmology, but models of galaxy formation are in good agreement with the observational data available so far.

A5.4 Cosmic microwave background anisotropies

A5.4.1 Statistical description of anisotropies

No discussion of structure formation would be complete without discussion of cosmic microwave background anisotropies, one of the fastest developing areas of astrophysics. The anisotropies in the cosmic microwave background are of particular interest because the temperature variations are so small, which means that the Universe was close to homogeneity when the microwave background formed and hence accurate calculations can be made using computer programs such as CMBFAST. Furthermore, it turns out that the predicted anisotropies are very sensitive to a wide range of cosmological parameters, meaning that accurate measurements of them can provide excellent constraints on cosmological models.

The fundamental measurement in microwave background studies is the temperature of the microwave background seen in a given direction on the sky, $T(\theta, \phi)$.[1] Usually the mean temperature $\bar{T}$ is subtracted and a dimensionless temperature anisotropy

$$\frac{\Delta T}{T}(\theta, \phi) = \frac{T(\theta, \phi) - \bar{T}}{\bar{T}}, \tag{A5.4}$$

is defined. The next step is to carry out an expansion in spherical harmonics $Y_m^\ell(\theta, \phi)$ (the analogue of a Fourier series for the surface of a sphere)

$$\frac{\Delta T}{T}(\theta, \phi) = \sum_{\ell=1}^{\infty} \sum_{m=-\ell}^{\ell} a_{\ell m} Y_m^\ell(\theta, \phi). \tag{A5.5}$$

[1]The radiation is also predicted to have a small level of polarization, and this was first detected in 2002 by the DASI experiment. Polarization can be described similarly to temperature, and is likely to become an increasingly important observational measurement.

The coefficients $a_{\ell m}$ tell us the size of the irregularities on different scales. As with the galaxy distribution, to compare with theory we are interested only in the statistical properties of these coefficients, quantified by the **radiation angular power spectrum**, now known universally by the notation C_ℓ and defined by

$$C_\ell = \langle |a_{\ell m}|^2 \rangle \,. \tag{A5.6}$$

This expression needs quite a bit of explanation. The angled brackets mean a statistical average, as often used in statistical mechanics. To a theorist that means an average over all the possible Universes that might have arisen by chance, once the cosmological model has been fixed. This can also be thought of as an average over all the possible observers in our Universe; remember that we just see the microwaves emitted from our own last-scattering surface, and other observers will see different photons whose detailed pattern of temperature variation may differ, just as observers in different galaxies would have a different view of the detailed galaxy distribution. However observers can only study the microwave background as seen from Earth, and all they can do is average over the different values of the index m. Care therefore needs to be taken in making comparisons with theory and observation; the difference between our region of the Universe as compared to the average region of the Universe is known as **cosmic variance**.

On average, a particular $a_{\ell m}$ is as likely to be negative as positive, and so the interesting quantity is the mean-square $|a_{\ell m}|^2$, which measures the typical deviation of $a_{\ell m}$ from zero and hence gives the typical size of anisotropies. The modulus signs are needed just because the usual convention is to define the spherical harmonics as complex and then impose reality conditions to ensure $\Delta T/T$ ends up being real. Finally, the requirement that the statistical properties are independent of the choice of origin of the θ–ϕ coordinates (rotational invariance) means that the result cannot depend on the m index, so the radiation angular power spectrum C_ℓ depends on ℓ alone.

The index ℓ can be thought of as giving the angular scale, with small ℓ corresponding to large angular scales and large ℓ to small angular scales. This is because as ℓ increases, the spherical harmonics have variation on smaller angular scales. As a rough rule of thumb, C_ℓ is telling us about the size of irregularities on an angular scale of approximately $180^o/\ell$. The interesting range for current observations runs from ℓ equals one up to ℓ of several thousand.

The most prominent feature in the cosmic microwave background is the $\ell = 1$ perturbation, known as the dipole. It corresponds to a pattern which is hot in one direction and cold in the opposite direction, with a smooth transition between them. It is believed to be due to the motion of the Earth relative to the microwave background, with the dipole simply due to the Doppler effect. Averaging over a year, its maximum value is $\Delta T/T = 1.23 \times 10^{-3}$, corresponding to the Sun having a velocity of $370\,\mathrm{km\,s^{-1}}$; taking into account the Sun's revolution around the galaxy this is consistent with the typical peculiar velocities observed for nearby galaxies. While interesting, this observation is not telling us about properties intrinsic to the microwave background, and so usually the dipole is studied separately and $\ell = 2$ is the smallest value considered. Maps of the cosmic microwave background are always shown with the dipole already removed.

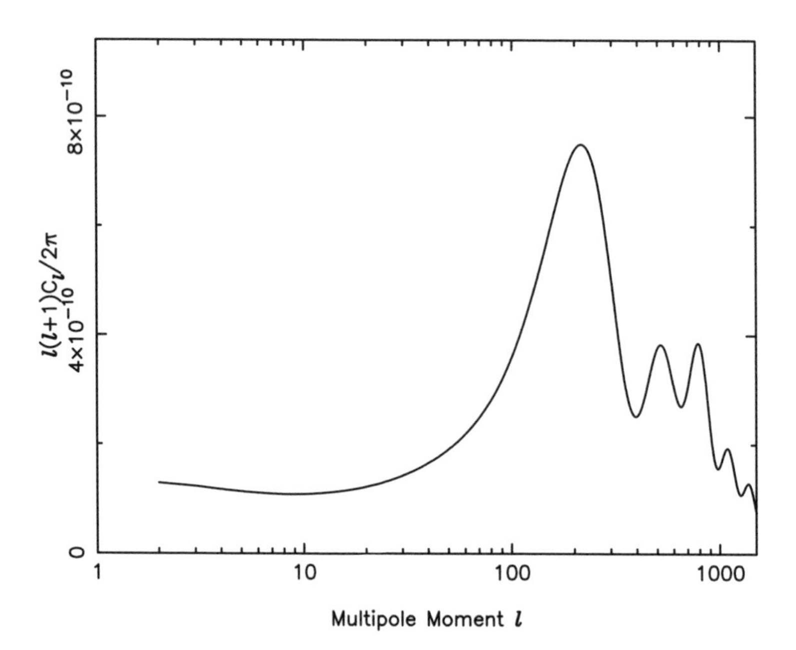

Figure A5.3 A typical prediction of cosmic microwave anisotropies, in this case for the Standard Cosmological Model. The predicted curve is calculated to better than one percent accuracy.

A5.4.2 Computing the C_ℓ

Because the anisotropies in the cosmic microwave background represent small departures from homogeneity, it is possible to calculate them accurately, though this requires a sophisticated numerical computation which includes many physical processes, such as gravitational attraction and the interaction of radiation with electrons. Before the microwave background is released, the photons are interacting strongly with the electrons, providing a pressure which opposes gravitational collapse. At that time, therefore, the cosmic fluid is undergoing oscillations, alternating between compression and rarefaction under the combined influence of gravitation and pressure.

At some stage in this process, the Universe cools sufficiently to release the microwave background. This removes the pressure support from the atoms, which are now able to collapse gravitationally to form galaxies and stars. But the photons are already on their way towards us, carrying a snapshot of the complicated fluid motions taking place at a redshift of around one thousand.

Modelling all these complexities is extremely challenging, but has been made accessible to the astronomy community via publically-available programs such as CMBFAST and CAMB. Such programs act as a 'black box', where you feed in your favourite cosmological model at one end, and out the other end comes a detailed prediction ready to be tested against observational data. For example, Figure A5.3 shows the prediction for the structure of microwave anisotropies in the Standard Cosmological Model.

As you can see from the figure, the predictions take on a complex form. By convention the combination $\ell(\ell+1)C_\ell/2\pi$ is plotted, and on the very largest scales, corresponding to small ℓ (for example COBE probed only $\ell \leq 15$), this has a non-zero value, with $\ell(\ell+1)C_\ell$ roughly constant across the COBE range. This region is known as the Sachs–Wolfe plateau, and is caused by variations in the gravitational potential between regions. Moving to smaller scales (larger ℓ) we see a broad peak at $\ell \simeq 200$ followed by a series of peaks and troughs which represent the complicated fluid motions that were taking place when the microwave background was released. Bearing in mind our rule-of-thumb for relating ℓ to angles, $\theta \sim 180^o/\ell$, the first broad peak at $\ell \simeq 200$ corresponds to an angular scale of about one degree, indicating that maps of the microwave background are predicted to have particularly strong features of that angular size.

Had I chosen a different cosmological model, the qualitative pattern of peaks and troughs would have been the same, but the detailed structure would change. Sufficiently accurate measurements of those structures can therefore rule out cosmological models.

A5.4.3 Microwave background observations

After the discovery of the anisotropies by the COBE satellite, which only probed the largest angular scales, most observers turned their attention to probing the structure predicted on smaller angular scales, requiring higher-resolution experiments. While many experiments contributed towards this goal, it is widely recognized that the landmark step was made in April 2000 with the announcement of results from the Boomerang experiment. This was an ingenious experiment carried around Antarctica on a high-altitude balloon by wind currents, in order to maximize observation time (around fourteen days) while minimizing contamination from the atmosphere. This experiment clearly picked out the first broad peak, locating it at $\ell \simeq 200$, the significance of which will be explored in the following subsection. It was rapidly followed by independent confirmation from another balloon experiment, called Maxima. Subsequently, more detailed analysis of these observations, along with results from new experiments, began to pick out the peak structures at larger ℓ.

Experiments carried out on Earth or on high-altitude balloons can have high precision, but their statistical power is limited by being able to survey only small areas of sky. In February 2003, spectacular results were announced by the WMAP satellite project. The true successor to COBE, this mission combined sensitive detectors with full sky coverage at a resolution approaching 10 arcminutes. The measurement of the radiation angular power spectrum based on their first five years of observations, released in 2008, is shown in Figure A5.4. A comparison with Figure A5.3 (noting the different ℓ-axis scaling) shows that these observations agree extremely well with expectations.

The solid line in Figure A5.4 is the prediction from the cosmological model best fitting their data. As discussed in Advanced Topic 6, this fit can now be taken as *defining* the parameters of the Standard Cosmological Model. The band indicates the statistical uncertainty from cosmic variance, which is most important on large angular scales. The conclusion is that our current cosmological models are very much on the right track, a remarkable vindication of theoretical predictions which were already in place long before any microwave anisotropies were measured. Those anisotropies are now the most powerful observational tool for constraining cosmological models.

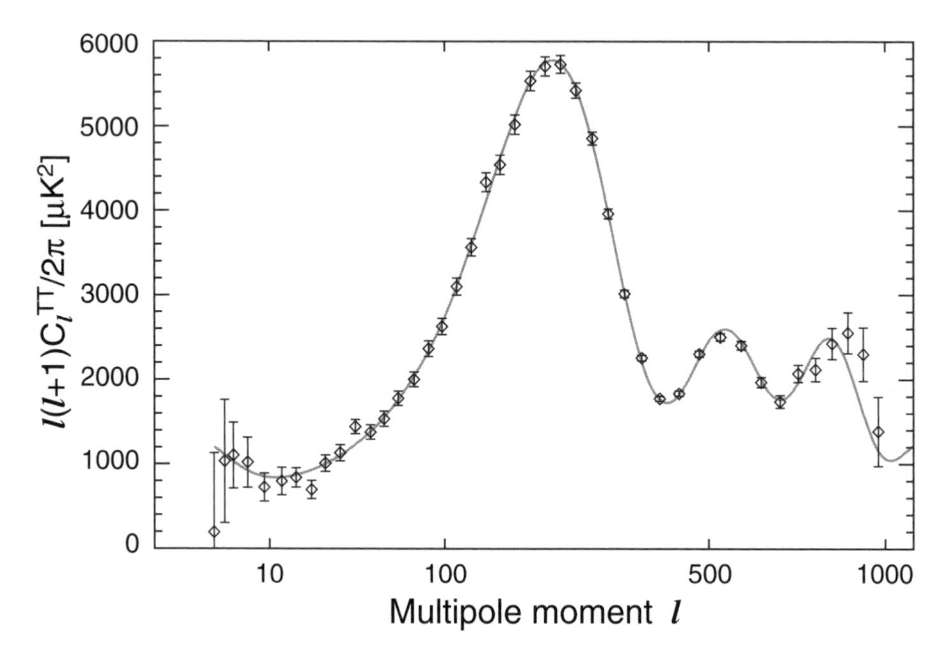

Figure A5.4 The radiation angular power spectrum as measured by the WMAP satellite from five years of data, shown as the black dots. The solid line shows a theoretical prediction from their best-fit cosmological model, which fits the data extremely well. They define C_ℓ using the multipoles of ΔT itself rather than $\Delta T/T$, so their scale is $T_0^2 = (2.725\,\mathrm{K})^2$ times that of Figure A5.3. Note that while Figure A5.3 uses a logarithmic ℓ-axis, this plot uses a non-standard scaling along the ℓ-axis chosen to display the observations evenly. [Figure courtesy NASA/WMAP Science Team.]

A5.4.4 Spatial geometry

While high-precision microwave background observations can constrain many cosmological parameters, one of the most important is that they give a direct indication of the geometry of the Universe, which can be read off from the location of the first peak in the angular power spectrum. While there is much complicated physics taking place around the time of formation of the microwave background, there is only one important characteristic scale, which is the Hubble time H^{-1} at that redshift. The peak structure comes from oscillations, and so the first peak, being on the largest scale, must correspond to perturbations which have just had time to undergo one oscillation. The Hubble time is an estimate of the age of the Universe at that time, and the Hubble length cH^{-1} estimates the physical size of a perturbation which oscillates on that timescale.[2]

In order to predict how the angular scale of the peak depends on our choice of cosmological model, we just need to know the Hubble length at last scattering and the angular diameter distance to the last-scattering surface. Problem A5.2 takes you through a demonstration that for spatially-flat Universes, the peak position is always approximately the same, independent of Ω_0. Problem A5.3 shows that if $\Lambda = 0$ there is a strong dependence

[2]This assumes that the sound speed c_s is approximately the speed of light; in fact it is a bit smaller, but close enough for my purposes.

on Ω_0. In combination, these indicate that it is the geometry which is the main factor in determining the peak location. These arguments are not however good enough to tell you the precise position of the peak, which needs a detailed calculation, but a glance at Figure A5.3 shows you that the peak is at $\ell \simeq 220$ for the Standard Cosmological Model, corresponding to an angular scale of about a degree. This is exactly where the WMAP results place the peak.

The WMAP satellite five-year results, combined with other observations, indicate that

$$\Omega_0 + \Omega_\Lambda = 0.995 \pm 0.013\,, \tag{A5.7}$$

placing the Universe within a percent or so of spatial flatness and completely ruling out significantly non-flat Universes. While this conclusion does depend on the validity of the assumptions made in the theoretical calculation of the C_ℓ, those predictions are borne out so well by the observations that discussion of non-flat Universes has become very rare. Figure A2.4 on page 132 shows the constraint in the Ω_0–Ω_Λ plane from the supernovae experiments. Only a tiny region, the location of the Standard Cosmological Model, can match both the microwave background and supernova data. Combining the microwave data with the preferred matter density of $\Omega_0 \simeq 0.3$ also gives support for the cosmological constant independent of the supernova results.

A5.5 The origin of structure

Gravitational instability is a powerful idea which lets us understand how structures in the Universe evolve. However, it does not let us address a more fundamental question — what is the origin of structure? Gravitational instability is excellent for taking initially small irregularities and amplifying them, but it needs the initial irregularities to act upon. Where might they come from?

The origin of structure takes us back into the realm of the very early Universe, because it appears that none of the established physics we know about is capable of making perturbations. However we do know of a mechanism that can. I introduced inflation in Chapter 13 following the historical motivation of the flatness and horizon problems. But in fact the best reason for believing in inflation (and certainly the best hope for testing the idea observationally) wasn't appreciated until a year after Guth's paper, and is that inflation can generate irregularities capable of initiating structure formation.

The mechanism is a remarkable one, being quantum mechanical in origin. Heisenberg's famous Uncertainty Principle tells us that even apparently empty space is a seething mass of quantum fluctuations, with particles continually popping in and out of existence. Normally we don't notice this as the time and length scales are so small, but during a period of inflation the Universe is expanding so rapidly that any fluctuations get caught up in the expansion and stretched. While one set of fluctuations is being stretched, new fluctuations are always being created which will then themselves be caught up in the expansion. By the end of inflation, there are small irregularities on a wide range of different length scales. Gravitational instability then acts on these small initial irregularities, and eventually, much much later, they can form galaxies and galaxy clusters.

This inflationary mechanism is currently the most popular model for the origin of structure, partly because it turns out to give mathematically simple predictions, but mainly because so far it offers excellent agreement with the real Universe, such as the microwave anisotropies just discussed. As I mentioned at the end of Section 13.5, there are presently quite a few different models for inflation, and typically their detailed predictions for the origin of structure are somewhat different. They therefore predict slightly different patterns of observed structures, hopefully different enough that one day we can use these structures to distinguish between inflation models observationally.

If the inflationary picture of the origin of structure is correct, a striking consequence is that all structures, including our own bodies, ultimately owe their existence to small quantum fluctuations occurring during the inflationary epoch. There can be no more dramatic example of the strong connection between microphysics and the large-scale Universe.

Problems

A5.1. You might wonder whether the galaxy distribution shown in Figure 2.2 on page 5 could arise by random chance. If you have access to a computer, you could try the following experiment. For the x and y coordinates of a point, get the computer to choose random numbers between 0 and 1. Repeat this until you have 1000 points (about the number of galaxies in Figure 2.2), and make a plot of them. Try this several times. Does the outcome ever resemble the real map?

A5.2. Throughout this question, assume that the Universe only contains matter and a possible cosmological constant. Use the definition of the density parameter to show that at any epoch we can write the matter density as

$$\frac{H^2\,\Omega_{\mathrm{mat}}(z)}{(1+z)^3} = \mathrm{constant} = \Omega_0 H_0^2\,.$$

Given that in any realistic cosmology $\Omega_{\mathrm{mat}} \simeq 1$ at early times, compute the Hubble parameter at decoupling ($z_{\mathrm{dec}} \simeq 1000$) as a fraction of its present value. As the only important characteristic scale in the young Universe, the Hubble length cH^{-1} gives the characteristic scale of the first peak in the microwave background.

In a spatially-flat cosmology with a cosmological constant, the present physical distance to an object with $z \gg 1$ is given approximately by

$$a_0 r_0 \simeq \frac{2cH_0^{-1}}{\sqrt{\Omega_0}}\,.$$

[If you wish, you can verify this approximation by numerical integration along the lines of Problem A2.4.] By considering the angular diameter distance, demonstrate that the angle subtended by the Hubble length at decoupling is approximately independent of Ω_0 in spatially-flat cosmologies, and compute its value in degrees. This demonstrates that the peak position is nearly independent of Ω_Λ for spatially-flat geometries, the approximations being the formula for $a_0 r_0$ above and the assumption that the Universe is perfectly matter dominated at last-scattering.

A5.3. This problem repeats Problem A5.2 for an open Universe with $\Lambda = 0$.

(a) *[For the mathematically-keen only!]* Show that in a matter-dominated open Universe with $\Lambda = 0$, the angular diameter distance to an object at redshift z is given by

$$d_{\rm diam} = 2cH_0^{-1} \frac{\Omega_0 z + (\Omega_0 - 2)\left(\sqrt{1+\Omega_0 z} - 1\right)}{\Omega_0^2 (1+z)^2} .$$

(b) Using the result of part (a) above, demonstrate that the angular size of the Hubble length at decoupling is approximately $\theta = 1\,\text{deg} \times \Omega_0^{1/2}$. Given that if $\Omega_0 = 1$ the peak in the microwave power spectrum is at $\ell \simeq 220$, use this result to predict the peak position in an open Universe with $\Omega_0 = 0.3$, and compare with Figure A5.4. This demonstrates that the peak position does depend significantly on geometry, and hence can strongly constrain it.

Advanced Topic 6

Constraining cosmological models

Prerequisites: Chapters 1 to 13 and Advanced Topics 2 and 5

A6.1 Cosmological models and parameters

Now that precision cosmological data exist, comparisons of cosmological models with data have become quite sophisticated.

Let's first be clear what a cosmological model actually is. It is a mathematical representation, intended to capture in sufficient detail the physical processes relevant to the types of observation we wish to describe. To be worth considering, the model must be consistent with what is already known about the Universe. To be useful, we must be able to ask our model to make predictions for the observations we plan to explain. Particularly valued are models that make specific predictions for observations yet to be made.

The first task is to define the physical processes governing the model, e.g. the types of material within the Universe, and the evolution equations such as the Friedmann equation. We may also have to specify initial conditions, for instance the type of density irregularities which initiate structure formation through gravitational instability. In order to understand the consequences of these equations, most likely we will have to model them on a computer.

Having decided on the basic principles underlying our model, it is likely that many quantities remain undecided, for instance the relative amounts of the different kinds of material in the Universe. These are the *cosmological parameters*, whose value cannot be predicted from first principles, but which we can seek to measure from our observations. If you like, the different possible values of the parameters describe the set of possible Universes consistent with the physical laws, but only one particular combination of parameter values describes our actual Universe and our aim is to seek which combination it is.

Consequently, we have a two-stage process. First, decide what the model is. Ideally this model should be as simple as possible, while sophisticated enough to explain the data we have obtained. Secondly, having decided the model, we use our observations to measure the values of its parameters, and in doing so learn about our Universe and its material composition. Ultimately, the hope would then be to exploit that knowledge to learn about the physical processes at work in our Universe, and perhaps connect them to ideas in fundamental theoretical physics such as superstring theory.

A6.2 Key cosmological observations

An impressive range of cosmological data now exists, and naturally the trend is for the data volume to continue a dramatic increase. One of the first tasks in a cosmological analysis is to decide which types of data should be used to test the model(s). Usually a compilation of several different types of data, analyzed simultaneously, is used. Desirable traits in a compilation of cosmological data include

1. The data should be high precision and reliable, and consistent with other quality data of the same type.

2. The data should be of a type where accurate and unambiguous predictions can be obtained from the models under investigation. Limiting factors here might include both the ability to model all necessary physical effects, and the availability of sufficient computing resource.

3. The observables should have a significant dependence on the model parameters of interest (otherwise the data will be unable to constrain them).

4. Where multiple datasets are used, they should complement each other, for example one type of data strongly constraining parameters (or parameter combinations) that other data constrain only weakly.

At present the most powerful single source of data is the cosmic microwave background anisotropies, as measured by the Wilkinson Microwave Anisotropy Probe (WMAP) (perhaps enhanced by other experiments of higher angular resolution). This scores powerfully under all of the criteria above, and indeed the simplest cosmological models can be well constrained using WMAP data alone. However, when more complicated models are considered, for instance allowing the spatial geometry to vary from flatness, WMAP alone is not quite good enough to constrain all parameters and is usually complemented by some additional data of other types.

A6.3 Cosmological data analysis

Cosmologists almost invariably tackle data analysis problems using a methodology known as Bayesian statistics, a system of inference named after the Reverend Thomas Bayes, who proved a key theorem, Bayes Theorem, in the mid eighteenth century. The Bayesian approach assigns a probability to each quantity of interest, for instance the probability of a parameter lying within a particular range of values, and then updates those probabilities in light of observations using a series of rules.

Bayes theorem states that given two possible statements/events, A and B, then

$$P(B|A) = \frac{P(A|B)P(B)}{P(A)} . \tag{A6.1}$$

where the vertical line indicates the conditional probability, usually read as 'The probability of B given A equals ...'. Here A and B could be anything at all, but we'll take A to be the set of data D and B to be the parameter values θ (where θ is a vector made up of the parameters being varied in the model under consideration), hence writing

$$P(\theta|D) = \frac{P(D|\theta)P(\theta)}{P(D)} . \tag{A6.2}$$

The left-hand side is what we want to know, namely the probability distribution of the parameters θ given the data, and this is referred to as the *posterior distribution*. On the right-hand side, $P(D)$ does not depend on the parameters and hence is just an overall normalization, and can ignored in estimating parameter values. $P(\theta)$, the *prior distribution*, is our knowledge of the possible parameter values before the data was obtained. Finally, $P(D|\theta)$, the probability of the data presuming particular parameter values, is known as the *likelihood*. It is this likelihood that we want to be able to calculate, so we can multiply our prior knowledge by it in order to update it to the posterior.

The above presupposes that we know which parameters θ we need to vary in order to fit the data. In reality we often do not, and indeed one of the primary goals of cosmology is to work out which physical processes are relevant to our Universe, and hence which parameters are needed to describe it. Bayesian inference can be extended to the situation where several different choices of parameters, i.e. models, need to be considered and compared. This extension is known as model selection, model comparison, or multi-model inference. However I shall not explore these ideas in this book, instead presuming that we do know which is the appropriate model to fit to the data and are focussed on determining its parameter values.

Computing the likelihood of particular parameter values requires calculating theoretical predictions from the model (for example, the C_ℓ curve describing cosmic microwave background anisotropies), and then working out how probable the observed data were given those theoretical predictions (essentially a glorified chi-squared analysis). The theoretical calculations can be carried out using publically-available computer packages, such as CMBFAST or CAMB described in Advanced Topic 5.4.2, and increasingly experimental teams are making available (as a piece of computer code) the *likelihood function* that combines those predictions with the data to generate the likelihood of the parameters.

The remaining challenge is to navigate around the space of possible parameter values in order to map out the posterior distribution of parameters. This is challenging, as typically many parameters, usually at least six, are being varied simultaneously, and calculating the likelihood at a single point in parameter space requires several seconds of computing time. Fully sampling the parameter space even quite crudely, say at ten points in each of six parameter directions, requires a million likelihood evaluations and hence the best part of a year of computer time, rather a long time to wait for results.

Fortunately, there is a much more efficient way of exploring parameter space, known as a Monte Carlo Markov Chain (MCMC). In this approach, the parameter space is explored by 'jumping' randomly from one set of parameter values to the next, with a rule for whether the jump is accepted or rejected depending on how the likelihood of the new point compares to the old. In the simplest version, known as the Metropolis–Hastings algorithm, jumps are always accepted if the likelihood of the new point is higher, while jumps to lower likelihood points are sometimes accepted and sometimes not according to a particular rule. The algorithm therefore generally drifts towards the highest likelihood

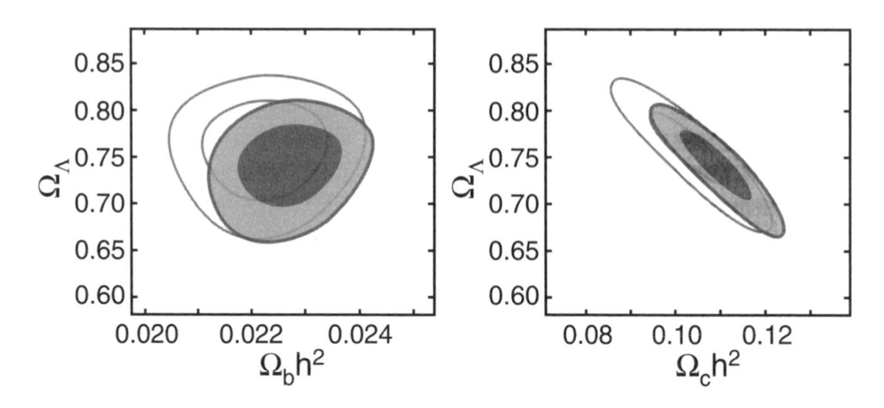

Figure A6.1 An example of joint constraints on two pairs of parameters, obtained from an MCMC calculation using data from the WMAP satellite. The filled contours, showing 68% and 95% confidence limits, are from the combined five-year dataset, while the unfilled contours show the results from the first thee years for comparison. The choice of variables on the axes of these plots is discussed in Advanced Topic 6.4; Ω_c is the density parameter for cold dark matter alone. [Image courtesy NASA/WMAP Science Team.]

regions, where the fit to the data is best, and then meanders around that region exploring the shape of the likelihood in the vicinity of the maximum. The Metropolis–Hastings rule is chosen so that this exploration maps out the probability of the parameter values.

Cosmologists are fortunate to have access to a software package CosmoMC, written by Antony Lewis and Sarah Bridle. This carries out an MCMC analysis using the CAMB program to carry out the theoretical calculations, and many datasets have been incorporated into it making it relatively easy for cosmologists to carry out analysis of new cosmological models or to discover the effect of emerging data. CosmoMC is a publically-available code,[1] but the reader should be warned that access to large amounts of computer time, typically on multi-processor platforms, is necessary to get reliable results from it.

Figure A6.1 shows an example of constraints obtained on two parameters using an MCMC calculation.

A6.4 The Standard Cosmological Model: 2008 edition

At the time of writing, the most powerful single dataset for constraining cosmological models is the five-year combined dataset from the WMAP satellite. Indeed, this dataset, usually referred to as WMAP5, is powerful enough that quite definitive results can be obtained using it on its own. However for highest precision some other datasets are added — measurements of supernova luminosities as described in Advanced Topic 2.3 and measures of galaxy correlations (in particular a phenomenon known as baryon acoustic oscillations in the galaxy power spectrum mentioned in Advanced Topic 5.3) — to give a data compilation known as WMAP5+all.

The data is well fit by models which assume spatial flatness. Six parameters need to be varied in fits to the data, and are as follows. First there are the densities of baryons, of cold

[1]Go to `http://cosmologist.info` for more information and downloads.

Table 6.1 Parameter values for the Standard Cosmological Model 2008. The model assumes spatial flatness, and allows six parameters to vary as shown in the upper part of the table. Here Ω_0 is the total matter density, i.e. the baryons and cold dark matter combined. The lower part shows other parameters that may be of interest which can be derived from this basic set. [Results, with some additional rounding, from papers by Dunkley et al. (2008) and Komatsu et al. (2008), both still to be published.]

Parameter	WMAP5 alone	WMAP5 + all
$\Omega_0 h^2$	0.133 ± 0.006	0.137 ± 0.004
$\Omega_B h^2$	0.0227 ± 0.0006	0.0226 ± 0.0006
Ω_Λ	0.74 ± 0.03	0.72 ± 0.02
n	0.963 ± 0.015	0.960 ± 0.014
τ	0.087 ± 0.017	0.084 ± 0.016
$\Delta^2_{\mathcal{R}}$	$(2.4 \pm 0.1) \times 10^{-9}$	$(2.5 \pm 0.1) \times 10^{-9}$
h	0.72 ± 0.03	0.701 ± 0.013
t_0	(13.7 ± 0.1) Gyr	(13.7 ± 0.1) Gyr

dark matter, and of the cosmological constant, along with the Hubble constant. Rather than fit these directly (with spatial flatness enforcing that the density parameters sum to one), it is standard to quote the baryon and dark matter densities as the density parameter multiplied by the Hubble constant squared. This combination is proportional to the physical density ρ, and it turns out that the observables are most sensitive to this combination. The spatial flatness condition then fixes h, which is no longer a free parameter in the analysis.

The remaining three parameters needed to get a good fit to the data are the amplitude and scale dependence of the primordial density perturbations, and the optical depth to reionization. These are indicated by the symbols $\Delta^2_{\mathcal{R}}$, n, and τ respectively, their precise meanings being beyond the scope of this book. The first two quantify the type of initial inhomogeneities, as may have been generated during the inflationary era. The last is a measure of the probability that a photon from the last-scattering surface collided with and scattered from a free electron on route to us.

Table 6.1 shows the current constraints on these six parameters, plus two auxiliary parameters that can be determined from them. Note the high degree of precision with which these are determined; for example, the age of the Universe t_0 is known to about one percent precision. The baryon, dark matter, and dark energy densities are also all measured to very good accuracy, and their values are illustrated in Figure A6.2.

Overall, these numbers give a more accurate representation of the Standard Cosmological Model than I gave in Chapter 15.

A6.5 The future

Future cosmological experiments have the dual goals of increasing the precision on already measured parameters, and on identifying a need for new parameters describing new physical processes at work in our Universe. In the absence of the latter, significant progress on precision would be expected, for instance from the Planck satellite which will make very high accuracy cosmic microwave background anisotropy measurements. This would be a welcome confirmation that our ideas of how the Universe works are broadly correct and

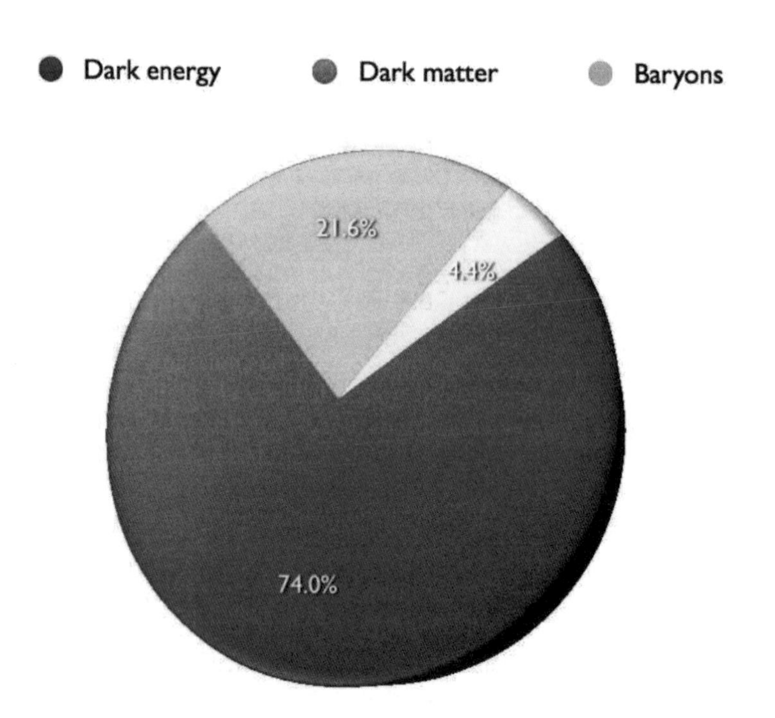

Figure A6.2 The three dominant constituents of the Universe. The present densities of photons and of neutrinos are too small to appear in this figure.

are fairly complete.

However, discovering new physical processes is what cosmologists really hope to get from new experiments, and most of us would be much happier to see something genuinely new. What form this might take is guesswork, but some example landmark discoveries that may lie in the future (from my perspective writing this) are

- Discovery that the cosmological constant is not, after all, a constant, but rather has a time-dependence to its density. This would indicate that some dynamical process is at work, and the cosmological 'constant' is then referred to under the more general name of *dark energy*.
- Discovery that the cosmic microwave background holds distinctive signatures of the inflationary cosmology (a particular hope is that gravitational waves — ripples in space-time — might be shown to have affected the cosmic microwave background).
- Discovery that the dark matter is not perfectly cold, but rather has some random motion which affects the details of galaxy formation and evolution.

If cosmologists like me are to stay in our job, we'd better hope at least one of these possibilities, or better still something completely unexpected, does indeed show up to give us new directions to explore!

Bibliography

A vast array of books on cosmology exist at all levels. Here is a far-from-complete selection. I would rate this book as Introductory Undergraduate.

Background

'The Origin of the Universe', J. D. Barrow, Orion, London, 1994,
'Before the Beginning', M. J. Rees, Simon & Schuster, London, 1997:
Very readable non-technical accounts of the Big Bang model.

Introductory Undergraduate

'The Big Bang', J. Silk, Freeman, 2001 (3rd ed):
Mostly text but quite detailed in many of the explanations. A classic text balanced between popular science and elementary undergraduate.

'Cosmology', M. Rowan-Robinson, Oxford University Press, 1996 (3rd ed):
A strong focus placed on observational aspects of cosmology.

'The Dynamic Cosmos', M. S. Madsen, Chapman and Hall, London, 1995:
Quite short, yet wide-ranging and accessible introduction to theoretical aspects of cosmology.

'Introduction to Cosmology', B. Ryden, Addison–Wesley, 2002:
Clearly written with strong use of analogies in explanations.

Advanced Undergraduate

'Introduction to Cosmology', M. Roos, Wiley, Chichester, 1997 (2nd ed):
A full account of the Big Bang cosmology, at a somewhat more advanced level than this book.

'Gravitation and Cosmology', S. Weinberg, Wiley, Chichester, 1972:
A classic textbook with exemplary presentation, largely focusing on general relativity but with strong material both on relativistic stars and cosmology.

Postgraduate

'The Early Universe', E. W. Kolb and M. S. Turner, Addison–Wesley, Redwood City, 1994 (paperback edition):
A detailed study of the dynamics and thermodynamics of the Big Bang cosmology, followed by material focusing on possible physics of the very early Universe.

'Principles of Physical Cosmology', P. J. E. Peebles, Princeton University Press, 1993:
An account of Big Bang cosmology, strong on historical detail, followed by highly detailed studies of the origin of structure in the Universe.

'Cosmology: the Origin and Evolution of Cosmic Structure', P. Coles and F. Lucchin, Wiley, Chichester, 2002 (2nd ed):
Again, a book primarily concerned with structure formation in the Universe, but including some material on the Big Bang cosmology itself.

'Cosmological Physics', J. A. Peacock, Cambridge University Press, 1999:
An extremely lucid account of many aspects of cosmology, both the Big Bang and structure formation, mostly at postgraduate level.

'Cosmological Inflation and Large-Scale Structure', A. R. Liddle and D. H. Lyth, Cambridge University Press, 2000:
I can't resist mentioning my other book, a highly-technical account of all aspects of the inflationary cosmology. However there is a significant gap in level between the end of this book and the start of that one, which one of the above books would be needed to fill.

Numerical answers and hints to problems

2.1: You should get something like $\rho \sim 10^{-26}\,\mathrm{kg\,m^{-3}}$ for the Universe. The Earth's density is about 10^{30} times greater.

2.2: Slightly different answers are possible depending how you deal with the rms velocity. You should get something like $r > 35\,\mathrm{Mpc}$ for $H_0 = 100\,\mathrm{km\,s^{-1}}$ and $r > 70\,\mathrm{Mpc}$ for $H_0 = 50\,\mathrm{km\,s^{-1}}$.

2.4: The frequency is $f = 3.3 \times 10^{15}\,\mathrm{Hz}$ and the temperature $T = 53\,000\,\mathrm{K}$.

2.5: The constant is $5.8 \times 10^{10}\,\mathrm{Hz\,K^{-1}}$. For the Sun, $f_{\mathrm{peak}} \simeq 3.4 \times 10^{14}\,\mathrm{Hz}$. When I wrote the question I was expecting this to be in the visible part of the electromagnetic spectrum (as the peak is when the spectrum is expressed in wavelength), but in fact the nonlinear transformation between f and λ shifts the peak into the near infrared.

2.6: For the microwave background, $f_{\mathrm{peak}} \simeq 1.6 \times 10^{11}\,\mathrm{Hz}$, and the corresponding wavelength $\lambda_{\mathrm{peak}} = 1.9 \times 10^{-3}\,\mathrm{m} = 0.19\,\mathrm{cm}$ [i.e. slightly more than 5 waves per centimetre, as in Figure 2.4]. The energy density is $\epsilon_{\mathrm{rad}} = 4.17 \times 10^{-14}\,\mathrm{J\,m^{-3}}$.

4.1: Begin by showing that the radius of the circle is $R \sin\theta$. At the equator $c = 4r$.

4.2: Consider a thin circular strip of width dr at radius r; the area will be the width of the strip times the circumference, and the number of galaxies in the strip will be the area times the density. You then need to integrate this expression from radius zero to radius r to get the total number of galaxies. Fewer galaxies are seen in the spherical geometry.

5.1: Energy is always, always, always conserved.

5.2: To apply equation (2.4) to photons, remember that their rest mass is zero.

5.3: For $0 < \gamma < 2$, the solution is

$$\rho(a) \propto a^{-3\gamma} \quad ; \quad a(t) \propto t^{2/3\gamma} \quad ; \quad \rho(t) \propto t^{-2} .$$

For $\gamma = 0$, this solution breaks down and is replaced by

$$\rho = \rho_0 \quad ; \quad a(t) \propto \exp\left(\sqrt{\frac{8\pi G \rho_0}{3}}\, t\right) .$$

5.4: We need $\gamma = 2/3$. Then $a(t) \propto t$.

5.5: Use the chain rule to convert time derivatives to θ-derivatives.

5.6: The solution is $a(t) \propto t$; $\rho(t) \propto t^{-3}$. Stable.

6.2: Look at the acceleration equation (3.18).

7.1: Respectively, they evolve as $1/a^4$, $1/a^3$, constant, and $1/a^2$. Radiation will dominate at early times, and the cosmological constant at late times.

7.2: To have a static Universe, we must have both $\dot{a} = 0$ *and* $\ddot{a} = 0$. The latter gives $\Lambda = 4\pi G\rho$, and then as both ρ and Λ are positive, a positive curvature is required in the Friedmann equation.

7.3: This is fairly straightforward using the generalization of the acceleration equation to include Λ.

7.4: This is solved by considering how the ratio of the densities evolves, and imposing the flatness constraint. The answer $\Omega \simeq 0.003$, $\Omega_\Lambda \simeq 0.997$. At late times the matter term can be ignored, and the solution is exponential expansion [c.f. last part of Problem 5.3]. At late times $q \to -1$.

7.5: Deriving this equation is fiddly. To obtain it, write down an equation for the ratio $\Omega(z)/\Omega_\Lambda(z)$ in terms of the present ratio, and impose the condition of spatial flatness. According to equation (7.11), acceleration began at $\Omega = 2/3$, corresponding to $z \simeq 0.67$.

8.1: $t_{\rm gal} = 6.6 \times 10^9$ yrs ; $h < 0.85$.

8.2: For the second part, use $x = 1 - \Omega_0$ as the expansion parameter.

8.3: Positive Λ gives acceleration, implying a smaller velocity at early times and thus that the Universe needed longer to expand to its present size.

9.1: $m_\nu = 28\, h^2$ eV. We need $h < 0.60$ with the numbers as given, and $h < 0.33$ with the improved number quoted in the question. See also Advanced Topic 3.

9.2: A typical estimate would be one thousandth of a parsec.

10.2: $T \simeq 2 \times 10^{10}$ K. The mass density at 1 sec was about $2 \times 10^9\,{\rm kg\,m^{-3}}$, about a million times that of water. The density matches water at $t \simeq 1000$ sec.

10.4: $n_{\rm e} = 2 \times 10^{17}\,{\rm m^{-3}}$; non-relativistic; $d \simeq 7.5 \times 10^{10}$ m. The interaction time is about 250 sec, much less than $t_{\rm univ}$.

10.5: You should find $T \simeq 5700$ K, confirming the result given in text.

10.6: The estimated radius is $2000h^{-1}\,\mathrm{Mpc}$. This underestimates (in fact by a factor three) because it neglects the expansion of the Universe during the light propagation. Answers to remainder depend a little on the assumptions you make concerning Ω_0 and the density of dark matter. The number of galaxies in the observable Universe is about 10^{11}; presumably it is just a coincidence that this is very similar to the number of stars in the galaxy. Using $\Omega_{\mathrm{B}} \sim 0.05$, the number of protons is about 10^{78}.

11.1: Taking everything into account, $\Omega_\nu/\Omega_{\mathrm{rad}} = 0.68$ [equation (11.1)]. See also Advanced Topic 3.

11.2: If you use the information from Problem 11.1 that the energy per neutrino is smaller by $\sqrt[3]{4/11}$, you'll find n_ν is almost identical to n_γ. Even if you don't it's still much the same. The estimate of neutrinos passing through your body can be done in several ways and I imagine has a range of answers. My estimate is that about 10^{16} neutrinos pass through you each second!

11.3: Solar temperature was achieved at a time of about 4×10^6 sec, while CERN energies were reached at $t \simeq 4 \times 10^{-10}$ sec, at a temperature of $10^{15}\,\mathrm{K}$.

11.4: $\Omega_{\mathrm{rad}}(t_{\mathrm{dec}}) \sim 0.04$ is a typical estimate.

12.1: Only hydrogen would form.

12.2: $Y_4 \simeq 0.025$ in this hypothetical Universe.

12.3: There are $8/9$ electrons per baryon.

12.4: I would say nucleosynthesis, but it's a rather subjective question!

13.1: Yes it can.

13.2: Inflation corresponds to $m > 1$. [If $m < 0$ the inflationary condition is also satisfied, but the Universe is contracting rather than expanding.]

13.3: $T = 3 \times 10^{25}\,\mathrm{K}$ is achieved at $t = 4 \times 10^{-31}$ sec. We then have $T \propto 1/t$, and reach $T = 3\,\mathrm{K}$ at $t = 4 \times 10^{-6}$ sec.

13.4: Light could have travelled about $3000\,\mathrm{Mpc}$ up to the present day. Up to decoupling, it could only travel $0.095\,\mathrm{Mpc}$, which is stretched by the subsequent expansion to $95\,\mathrm{Mpc}$. The subtended angle is 2°.

13.5: The densities are equal at $T = 3 \times 10^{18}\,\mathrm{K}$. Today, $\Omega_{\mathrm{mon}}/\Omega_{\mathrm{rad}}$ would be 10^{18}.

13.6: An expansion factor of about 10^6 is required.

14.1: $T_{\mathrm{Pl}} \simeq 8 \times 10^{31}\,\mathrm{K}$.

A1.1: r has range $0 \leq r \leq 1/\sqrt{k}$. The equivalent transformation for the hyperbolic case is $r = (1/\sqrt{|k|})\,\sinh(\sqrt{|k|}\,\xi)$. The ratio of circumference to radius at $\xi = 10/\sqrt{|k|}$ is about 6920, rather than the usual 2π.

A1.2: The maximum physical separation is $s = \pi a_0/2\sqrt{k}$.

A2.1: The light is redshifted.

A2.2: The luminosity distance is given by

$$d_{\rm lum} = \frac{a_0}{\sqrt{k}}\,(1+z)\,\sin\left(\frac{\sqrt{k}}{a_0}\,d_{\rm phys}\right).$$

For nearby objects the sin can be expanded for small argument. For distant objects, redshift increases $d_{\rm lum}$ as compared to $d_{\rm phys}$, but the geometry, expressed by the sin function, acts to reduce $d_{\rm lum}$. This latter effect is because with spherical geometry the area at a given radius is smaller than would be given by the flat geometry, so the flux per unit area must be higher. The geometry could be thought of as focussing the light rays.

A2.3: An object with physical size l subtends an angle

$$\theta = \frac{l}{3ct_0}\,\frac{(1+z)^{3/2}}{(1+z)^{1/2}-1}$$

For small z we have $\theta \propto 1/z$ and for large z we have $\theta \propto z$.

A2.5: The number of sources scales as $N(>S) \propto S^{-3/2}$. As the number increases sharply as S is decreased, we conclude that most sources are seen at close to the flux limit.

A3.1: The cross-section is $\sigma \simeq G_{\rm F}^2 k_{\rm B}^2 T^2$. The interaction rate is $\Gamma = n\sigma v$ where $v \simeq c$ is the velocity. Putting these together gives the result.

A3.2: The effective number of species before annihilation is $g_* = 2 + 4 \times 7/8 = 11/2$, and afterwards is 2. Conservation of $g_* T^3$ across the transition gives the result.

A3.3: The redshift is $1 + z_{\rm nr} \simeq m_\nu c^2/3k_{\rm B}T$. For a 10 eV neutrino this gives $z_{\rm nr} \simeq 20\,000$. The comoving distance travelled by those neutrinos is approximately 8 Mpc. (If your answer is much smaller than this, you may have forgotten to allow for the expansion of the Universe *after* the neutrinos become non-relativistic.)

A5.1: Your randomly-generated maps won't look like the real thing. The strong filaments and large voids of the original won't be reproduced. Galaxies exhibit stronger clustering than a random distribution.

A5.2: This problem, though more sophisticated, is closely related to Problem 13.4. The Hubble parameter at decoupling is given by $H = \Omega_0^{1/2}(1+z_{\rm dec})^{3/2}\,H_0$. Using the angular diameter distance formula gives an apparent size $\theta = 1/2\sqrt{1+z_{\rm dec}} = 0.016\,{\rm rad} \simeq 1^o$.

A5.3: Part (a) is a nasty algebraic slog. For part (b), the expression for the Hubble radius at decoupling in Problem A5.2 remains valid, and we only need the $z \gg 1$ limit of $d_{\rm diam}$. For $\Omega_0 = 0.3$, we predict $\ell_{\rm peak} \simeq 400$, completely incompatible with the data.

Index